A Sonoma County Phenomenon:
Evidence of an Interdimensional gateway
By Margie Kay

A Sonoma County Phenomenon:
Evidence of an Interdimensional Gateway
By Margie Kay

Published by Un-X Media
PO Box 1166
Independence, Missouri 64051

Printed in the USA

ISBN: 978-0-9988558-5-1

All rights reserved. No portion of this book may be reprinted in any form whatsoever including print or electronic means without express written permission from the author.

©Copyright 2020 by Margie Kay

All photos copyright 2020 by Natalie Roberts and Margie Kay
Some photos obtained from Google Earth and Adobe Stock

Contact the publisher: editor@unxmedia.com

Acknowledgements

Thank you to Art Campbell for introducing me to Natalie Roberts and requesting that I investigate this case.

Thank you to Natalie Roberts for taking the time to provide photographs and information about the phenomena at her home, and for allowing me to conduct my own investigation.

Thank you to Wayne Lawrence, for analyzing several of the photographs provided by Natalie in order to determine their authenticity.

Thank you to the late Stanton Friedman, for encouraging me to investigate this location. I hope you are watching Stanton!

Table of Contents

Chapter	Page
Introduction	6
The Backstory	7
Remote Viewing and Investigation	15
The Light Beings: Photos of the Phenomena	18
Apparitions	59
Alien Craft	65
Planets and Suns	69
Natalie's Favorites	79
Our Multi-Dimensional Universe	97
Conclusion	99
References	101
About the Author	102
Publications by Un-X Media	103

Introduction

Sonoma County, California is known for its wine country and breathtaking views of the ocean. But it is also the site of one of the strangest phenomenon in the United States, if not the world. This book is a compilation of some of the best photographs ever obtained of a portal, and possibly a worm hole located at an undisclosed location in California. The photographer, Natalie Roberts, has been taking photographs of these strange anomalies in her back yard since August 16, 2006. I have been investigating this case since 2011, after my friend UFO investigator and author Art Campbell sked me to speak to Natalie and possibly remote-view her location to help determine what is going on at this site.

After contacting Natalie to determine what she has been experiencing, I found out that ufologists and scientific investigators have been to this location several times, and that they, too, find the photographs Natalie has taken to be outstanding and perplexing. One such person is Jacques Vallee. Jacques is a computer scientist, venture capitalist, author, ufologist and astronomer currently residing in San Francisco, California. He is well-known among ufologists for his scientific approach to UFO investigations, and if Jacques Vallee is interested in this site, it is something of which to take note. Natalie has taken over 250,000 photos to date using 16 different cameras both with and without flash. She usually takes the pictures at night after the sun goes down, as that is the best time to capture objects on film at this location. It is likely that events happen 24 hours a day, but due to the sunlight the objects don't usually appear on film during daytime hours, as is typical.

No one knows yet if the objects she has captured are extraterrestrial craft or beings, spirits, inter-dimensional beings, ghosts or as bizarre as this sounds – galaxies. Some of the objects do resemble tiny galaxies and stars. In order to make sense of the phenomena, I decided to approach this case like any other and start at the beginning, when Natalie was born. I always like to know a person's history- to see if there is something ongoing that they experience which may explain more about their experiences. It is interesting to note that in most paranormal cases, the subject has had a lifetime of strange experiences.

After communicating with Natalie for several years and examining her photographs, I've come to the conclusion that she has captured genuine unexplained phenomena at her home in Sonoma County, California, and that the public deserves to know about this phenomena and see at least a few of the many pictures of these objects. I hope the reader enjoys this book.

Margie Kay

The Backstory

Natale Roberts and her identical twin sister were born two months premature at the Lettermen Hospital in the Presidio, San Francisco in 1955. The Presidio is a park near the Golden Gate Bridge, but was once a military installation. Natalie's father worked there after he lost his leg in battle in WWII. He was a marine Captain in charge of personnel. Natalie's father made several trips to Washington, DC where he met with President Johnson regarding the work he was doing.

Natalie's father didn't speak much about his work, but he did tell her a couple of very strange things that stuck with her. It seems that there were quite a few men who went missing off of vessels that traveled through the Bermuda Triangle area, and he was the one who had to inform their families of their disappearances. It is quite well known that the Bermuda Triangle is a mysterious location, with many ships and planes reported missing over the years, but this is the first time I'd heard of people actually missing off of vessels that were unaffected. And for there to have been "quite a few" is an astounding revelation.

The second thing that Natalie's father told her that she found amazing was during a UFO sighting they both had at her parent's home in Mill Valley, California at the base of Mount Tamalpias. Natalie has fond memories of this home, and having a wonderful childhood there. But one night Natalie drove to the house when she was 21 years old to visit her parents something happened that would change her life forever. As she parked her car in the driveway, she noticed some very bright lights above the cedar trees next to her house. Her father appeared in the bedroom window and yelled "Get in here quick - they're here!" Natalie ran to the house while replying "What do you mean - They're here?" Her father said the object was a UFO. They both watched the very large triangular-shaped UFO with blinding lights on the bottom that lit up the yard move silently and slowly through the neighborhood for several minutes, then suddenly shoot away at incredible speed.

Natalie was struck by the way her father reacted- as if he knew all about UFOs and this was not an unusual occurrence. However, Natalie did not question her father further, which is something she regrets to this day. When her father got older, he did say something to Natalie that surprised her – he said for her never to trust the government – whatever they say is not true. She didn't get any further explanation, but heeded his warning. Natalie's father passed away in 1985 at age 66 due to colon cancer. Towards the end of his life, he kept writing and saying the word "Always," but no one knows why. Shortly after her father's death, the movie "Always" starring Richard Dreyfus came out. She does not know if there is a connection, but I find it odd that Richard Dreyfus starred in the biggest UFO movie of all time (Close Encounters), and that this family has had personal experiences with UFOs. I wonder what else Natalie's father knew about UFOs, and what he discussed with President Johnson, but we'll never know.

What Natalie didn't know at the time, was that her life would get very complicated when she got older.

Family

Natale married Barrett Roberts in 1980, and they had three children, and now have five grandchildren. Most of the children or grandchildren have not had any unusual experiences, however, Natalie's oldest daughter saw a UFO one time, and she also sees the strange objects in Natalie's back yard with the naked eye. The others rarely see the orbs and other objects with the naked eye – the anomalous objects only appear on camera most of the time. One other strange thing to note is that even if both women take photos at the same time, there will always be more objects in Natalie's photos than her daughter's.

This may be a form of Psychophotography – where a person is in telepathic communication with beings, and the beings or intelligences will allow themselves to be photographed by that person but not others. In Natalie's case, since she's been aware of the phenomena for years and has purposefully tried to communicate with it/them, and been the one who has taken the most photographs, it is not surprising that this type of photographic phenomena would occur.

First Contact

Natalie first noticed that something strange was going on at her house when her husband called her into the bedroom one afternoon in 1994 while he was napping after a long night at work. The family had just moved into the house. He pointed to a small light (orb) that was moving about the room. They both watched as the orb then went out the window and over to a tree in the back yard. They both noticed balls of colored lights around the tree and wondered what they were. After that, no one in the house noticed anything unusual until one day in 2006 when Natalie suddenly had the thought to take pictures at night, even though she didn't know if the photos would come out.

Natalie asked her daughter to stand in the back yard in front of the Apple tree one evening so she could take a picture of her using flash with her brand new digital camera. Natalie snapped a photo, and was surprised to see that throughout the apple tree, branch by branch, there were glowing balls that resembled a string of Christmas lights. She thought, "It's not Christmas, and besides, I never put lights in our apple tree. What the heck is going on?" Natalie was confused and bewildered by this situation. The light globes were in all sizes- large, small, and medium, with some brighter than others. It reminded her of the balls of light she saw at the tree years earlier.

Natalie then cleaned off the lens of the camera just in case there were smudges on it and took another picture. The results were the same, but the lights had moved to different positions.

Natalie took her camera to a camera repair shop to see if there was something wrong with it that would cause the strange lights to appear in a photo, but not to the naked eye, but the camera was working property.

The next day, Natalie took photos in the daylight, but nothing anomalous appeared. She decided to take more pictures at night to see if the strange lights would appear again. She thought it unlikely, but figured it would be worth taking a few pictures to find out, and she was surprised to find out that the lights, or orbs, appeared in nearly every picture. She photographed a large selection of objects including orbs, orbs with designs, smoky figures, and rod-looking shapes with lights inside of them.

Natalie started taking pictures of the tree both day and night. At the time, neither Natalie or her husband knew what an orb was and had to do research to find out what it might have been.

Natalie began taking photos of all locations in her yard and of the night sky and found that the objects appeared in many locations, not just at the apple tree. Natalie became somewhat obsessed when she was able to capture many strange objects in the photos, and decided to document what she was seeing on a nightly basis. Natalie has had no training in paranormal phenomena, so she reached out to me and other investigators in order to try to determine what is happening at her home.

Natalie Roberts in her back yard
Photo: Natalie Roberts

When Natalie initially began taking pictures her hands would hurt, and she got sick to her stomach quite often. This gradually dissipated, but there is no explanation. Coincidentally, the images were blurred and fuzzy, but gradually become more clear as time went on.

Natalie feels that the frequency (vibration) of the phenomena was something that she had to become accustomed to, and that in turn, the phenomena became accustomed to her. In other words, It (the phenomena) perhaps *allowed* her to take photographs.

The Experiments
Natalie shared her photos with friends and family, who were skeptical. They thought that the objects in the photos were caused by rain, moisture, smudges on the lens, or other explainable artifact. So they put a sprinkler on the roof to simulate rain, sprayed a mist in the air, and blew smoke in front of the lens, but none of these techniques produced anything like the objects in Natalie's pictures.

When Natalie resumed taking photographs she noticed that sometimes she could see the beings or objects with the naked eye just before a photo was taken. Sometimes the objects appear just out of the corner of her eye, but disappeared when she tried to look directly at them. Once she saw a large shape floating over her neighbor's rooftop headed towards her yard. Natalie feels that she has adjusted to the frequency of these objects and that is why she can now see them.

Note: Natalie has no doubt unknowingly learned how to use her night vision and third eye, which is what psychics use to see inter-dimensional beings.

Natalie feels that at least some of these objects are beings of light. They seemed to be attracted to the camera flash, and flocked to it. At first, just a few beings would approach, but later they would flock together in large numbers. Natalie is convinced that these beings were intrigued by the light from her camera.

One evening Natalie was standing on her patio getting ready to snap a photo when her flash reflected off something large right in front of them. They both jumped back at the same time and exclaimed, "Did you see that?" They ran inside the house and examined the photos. The camera had taken three pictures and in each, the object moved upward. The next day, Natalie's husband said that the object looked like something he found on the internet about Rod Ships in Mexico, and the description matched the object in her photos.

She decided to go through her photos taken previously to see if there were any matching the rod ships and there were quite a few. Some had lines going from the ship down into the ground, back up into the ship, then straight up into the sky. It appeared as if these objects might have been sucking something out of the ground and pumping it up to an unknown location.

The only things underground in Natalie's yard that she knows of is a leech line and septic tank, and she thought that might be something they use for fuel. Natalie checked her yard and found small burn marks in the ground, similar to cigarette burns, all over the lawn. There is no mundane explanation for this.

Shadows appear on the ground in some of the photographs, indicating that some physical object must be there to cause it.

At times, Natalie has experienced a change in temperature around her, or a cold, ectoplasm-type smoky mist enveloping her hands as she holds the camera. An investigator found that there were temperature differences in several locations in the yard, but could find no explanation.

Natalie feels that the visitors are attracted to water as some of her most colorful and detailed pictures have been taken during rain showers.

The Pig Man/Alien

I needed more information, so asked Natalie if she remembered anything strange that happened to her as a child. Indeed, there was something bizarre. She has had ongoing nightmares since childhood about a very ugly half pig, half humanoid alien appearing in from outside her bedroom window at her parent's home. The window was located three stories from the ground, so this entity must have been levitating. He would laugh at her, then come in the room through the window and make her go with him to another unknown location. Natalie clearly remembers this being telling her not to use her mouth to communicate, but rather to use her brain, and he would actually hold his hand over her mouth to keep her from talking.

Natalie's twin sister recently recalled seeing this same entity in their house, and remembered seeing him fly out through the bedroom window when he left them. At this point, Natalie had the realization that that what she thought was a recurring dream may not have been a dream at all. Since she was a child, Natalie was extremely afraid of the dark because of this entity– until 2018.

One evening in 2018 at approximately 11:00 pm, Natalie was sitting up in her bed. Her husband was at work and she was alone in the house. She had suddenly become very afraid of the dark again and was afraid to go to sleep, but she believes that she somehow must have fallen asleep anyway. Suddenly, a human shape, brightly light up with electrical rainbow lights all over him, floated down through her skylight and threw her covers off of the bed. It was the pig entity again. He also told her that wanted her to know that he is watching over her and not to worry. She asked if it was OK for her to take photos of the objects in her yard and he said yes. This calmed Natalie down, and for some reason she is no longer afraid of this being or of the dark.

Natalie awoke suddenly after this encounter to find all the lights on and her covers in a corner across the room. For this reason, Natalie does not believe that this was a dream and that she was actually visited by a living

A Sonoma County Phenomenon

Above: Natalie's story and several photos were first published by me in Un-X News Magazine in the Spring of 2012, not long after our first conversation. After I saw the pictures I knew that this location warranted a full investigation.

three-dimensional entity. One thing of note: Everyone in the family has said that they feel they are being watched at one time or another. The grandchildren have seen shadow people, and one in particular wears muddy boots. Shadow people are three-dimensional human-shaped beings with no features. They've seen this shadow person more often than the others. The family has heard someone walking on the sidewalk in the back of their house, but when they investigate no one is there. Sometimes the water spigot for the hose comes on right after they hear the sound of a person walking.

The Aliens

Recently, Natalie visited a psychic fair. As she strolled around the booths a man she didn't know walked up to her and said, "I just want you to know that when you walked into the room, all of the aliens in the room want over to you and hung around you and floated above you on the celling." The man then turned and walked away. Natalie was taken aback, but after thinking about it decided that the man could have been correct. Perhaps he was psychic and really saw what most others do not. After all of her experiences in the past, and her current connection to the vortex in her yard, Natalie believes that it is entirely possible that extraterrestrials are drawn to her.

The Water Mystery

Natalie and her neighbors have had some strange things happen regarding water. This began about one year ago. The taps in their kitchens and bathrooms often come on of their own accord and at full blast. This occurs at different times of the day and night, and even when they are not home- leaving a big flooded mess at times. Two of her neighbors have had the same thing happen. Plumbers have inspected the faucets and cannot figure out why this is happening. I believe it has something to do with the unusual energy at this location.

Additionally, two neighbors told her independently that they have seen "people" or shadow people walking around in their houses at night, then they disappear. No one is ever found in the houses. This could be another sign of a portal.

Ley Lines and Vortex

A portal or vortex is a location where there is a large amount of energy caused by underground water, ley lines, or perhaps a crossing of lines of latitude and longitude. Strange phenomena seem to happen at these types of locations more frequently than others. Some people are affected when they are near such a location, either positively or negatively. I've investigated a number of so-called "paranormal hot spots," which are frequently located near such crossings. For example, at 94 degrees longitude and 39, 38, 37, and 36 degrees longitude in Missouri and Arkansas there are frequent sightings of UFOs, orbs, and spirits, and even Sasquatch. There is often paranormal activity at these locations as well. The number of reports is staggering, and impossible to ignore. Something is facilitating the movement of entities and craft through dimensions at these particular locations, and the one thing they have in common is their close proximity to the lines of latitude and longitude. It is interesting to note that Natalie's home is located very close to 38 degrees latitude, which has been a focus of my investigations for many years. I don't believe in coincidences.

Some of the cameras Natalie has used to to take photographs:

1. Cannon power-shot 4.0 mega
2. Olympus Camedia D435 5.1 mega
3. Olympus 4.0
4. Cannon power-shot 6.0 D630
5. Sony cybershot 10.0 mega
6. Sony cybershot 12.1 mega
7. Nikon coolpix 12.1 mega
8. Nikon coolpix 14.1
9. Samsung 3515 10.2 mega
10. Lumix model dmctz4 8.0 mega
11. Rebel Cannon
12. Cannon powershot sx30IS 14.1

Ley Lines are lines of power which can extend over many miles. All of the megalithic structures around the world are located on Leys. In Ray Parkes blog at https://rayparkes.com/California-ley-lines, he discusses the major Ley Line intersections, which coincidentally have a central location at San Francisco, which is just south of Natalie's house. The leys literally cover the area where she lives. It is my opinion that the combination of the powerful ley lines and the crossing of 38 degrees longitude and 112 degrees latitude is making a very powerful vortex area, which may explain the unusual phenomena in the region.

A Worm Hole?

In the spring of 2019 Natale picked her husband up at his place of work at 3:00 am and headed home. They were watching the red-colored sky and smoke under the full moon. The red color and smoke were caused by fires in the area. While looking outside her

This drawing from Adobestock.com is similar to what I saw during a Remote –Viewing session to see what is happening in Natalie's back yard, except that I also saw stars, planets, and entire galaxies going in or out of this worm hole.

passenger side window, Natalie saw a banana shaped UFO fully lit up with bright lights. The object flipped up on its side and appeared as a disk shape. The craft then shot upwards a few seconds later at incredible lightning speed. Unfortunately, Natalie's husband did not get to see this object. The two did see a large black hole in the sky where the craft disappeared, and it was located right above their house. This hole was visible until the next morning, then it finally disappeared. Natalie believes this was a worm hole opening and closing, and so do I.

Remote-Viewing and Investigation

I remote-viewed Natalie's back yard at least 20 times in an effort to see what is going on there. RV is a method I use to see locations anywhere. It requires the use of the sixth sense and an activated pineal gland. I discuss the method in my book *The Remote Viewing Workbook*, and teach this method in seminars. Essentially, I meditate, then send my consciousness out to a location. I can then see what is happening at that site. It doesn't matter when an event occurred, so I'm able to look at past or present events.

I've used this technique to help solve over 60 missing person, homicide, and theft cases for law enforcement, in countless paranormal and UFO investigations, and have helped other investigators as well. I had remote-viewed some artifacts for author and ufologist Art Campbell in 2010 and the results were very accurate, so this is why he wanted me to get involved in this particular case.

What I saw during most of my RV sessions at Natalie's house is a circular area which is usually approximately 25' in diameter, and extends upward into the sky, and also underneath the earth. This size changes sometimes and extends outward to approximately 100'. Each time I look, different objects appear, but also there is a single consciousness involved, and at other times there are multiple intelligent entities. I've seen some faces, but most of the time I see bright objects of different shapes and sizes.

When remote-viewing this site one day I saw a large worm hole which goes all the way through the planet to the other side and out into space for some distance. I saw craft going in to the worm hole on the right side and out of it on the left side. I was in contact with a consciousness that explained this is how craft move through a worm hole, and they can go anywhere in the solar system, or to and from any place in our galaxy and to other dimensions using this method. When the craft enters the worm hole the occupants or the pilot simply has to think about where they want to go and they are almost instantaneously there. I decided to ask about how long it takes to get from here to Mars, and was told that most craft go to the moon first, then go to another worm hole to get to Mars in seven minutes. This part was not too surprising, but what I would hear and see next during this session was shocking.

During the next RV session of Natalie's back yard I saw what looked like stars, planets, and even entire tiny galaxies moving through this worm hole to other dimensions. I also heard a voice explaining this to me. Apparently, the celestial objects are their normal size to start with, but must shrink down to a tiny size in order to move through the worm hole, then are restored to their original size when they reach their destination on the other side. Any occupants are unaware of the transition. The "other side" is another dimension. Needless to say, I was quite shocked by this development as I'd never contemplated this.

Right after this session I called my friend nuclear physicist Stanton Friedman, and asked what he thought about the possibility that this could exist. He said he thought it certainly was a possibility, and that it should be further investigated.

When I say "hear," I mean that some type of consciousness gave me this information, but I don't know if it was an individual, or a collective consciousness such as the Akashic Record. How this could be proven I don't know, but perhaps some enterprising scientists have an idea. I have no training in science, so can only report on what I see and hear and let someone else try to figure it out scientifically. The fact that Natalie and her husband saw the black hole in the sky, and that there are thousands of strange objects or beings passing through their back yard is possible evidence that a worm hole does exist at this location, hence the name *A Sonoma County Phenomenon*.

My next step was to send five of Natalie's photos to Wayne Lawrence for analysis. Wayne is a UFO investigator for MUFON and a photographic and video analyst, and has looked at many photos for me for other cases. I completely trust his opinion. He found no manipulation of the photos, meaning that they are in their original form and not "Photoshopped."

I then decided to investigate past sighting reports on reporting websites. On MUFON.com, there were 179 reports of UFOs in Sonoma County, California from January, 2001 to June of 2019. Of note is that most of these reports are of bright lights, and in many cases, multi-colored lights at low altitude, which is not unlike what Natalie has photographed.

I checked with the National UFO Reporting Center, and there are over 13,000 UFO reports filed in California, so I didn't go through each file to find sightings in Sonoma county. I did find a few which were surprisingly similar to the MUFON reports.

Next, I did an online search for paranormal activity in Sonoma County, thinking that there might be similar types of things happening to other people in the area that would corroborate Natalie's experiences.

I found a website about paranormal activity along hiking trails: https://backpackerverse.com/10-sonoma-county-hiking-trails-with-insane-paranormal-activity/5_West_County_Regional_TrailSebastopol. This is some really strange stuff—you rarely hear about paranormal events on hiking trails. Of note are reports of shadow people along Pomo Canyon Trail that amazingly attach themselves to people's backs. I'd be very watchful if hiking this trail. It is interesting that these events are occurring at this particular location.

Here is another article online at https://www.onlyinyourstate.com/northern-california/strange-phenomenon-norcal/ in which the author discusses Gravity Hill in Lichau Road, where cars put in neutral actually roll uphill! Amazingly, gravity anomalies have been associated with worm holes by some investigators.

Ray Parkes has studied the World Grid and in particular, the Ley Line crossings in California. I contacted Ray by email and he said that the town of Sebastopol, where Natalie lives is located inside the largest lines that he calls the "Rods of Power." Ray also says that these main arterial lines are the life force of the planet. Ley Lines or energy lines can be detected by dowsing. Ray's blog is worth a visit for more information: https://rayparkes.com/california-ley-lines/

I've been studying the World Grid and Ley Lines for many years, focusing on the Midwest region of the U.S. One such power center happens to be where I live at Kansas City, and we experience many strange and unexplained things at this location. K.C. is one of the biggest hot spots for paranormal activity in the world, and I believe it has something to do with the Ley Line crossings and the intersection of 39 degrees latitude and 94 degrees longitude and the Ley Line crossing. See my book *The Kansas City UFO Flaps* for more information.

I've also investigated multiple locations at 38, 37, and 36 degrees latitude at 94 degrees longitude and find similar events happening at these locations as well. One such hot spot is the Board Camp Crystal Mine south of Mena, Arkansas. This site is known for levitating rocks, UFO sightings, Sasquatch encounters, and more. It sits nearly perfectly on 36 degrees Latitude and 94 degrees Longitude. If it is true that intersecting coordinates and ley lines cause portals, then there should also be a lot of paranormal activity going on in Sonoma County, California, and apparently there is.

I invite the reader to review the photos following and come to their own conclusions.

The Light Beings

The following photographs were taken by Natalie Roberts from 2006 to 2019 and by my night vision hunter camera in 2018. The photos were taken from Natalie's back yard and her house. None of the pictures were taken through a glass door or window, so there are no reflections. Most pictures were taken during winter months during dry weather, when it seems that the objects are easier to see and photograph than at other times of the year.

See what you think.

Photo #1:

Right:
Enlargement from photo above

A Sonoma County Phenomenon

Photo #2

Right: Enlargement of object at the bottom of the photo, just above the house. This object appears in several photos. It may be a planet—but why are there foggy lines in front of it?

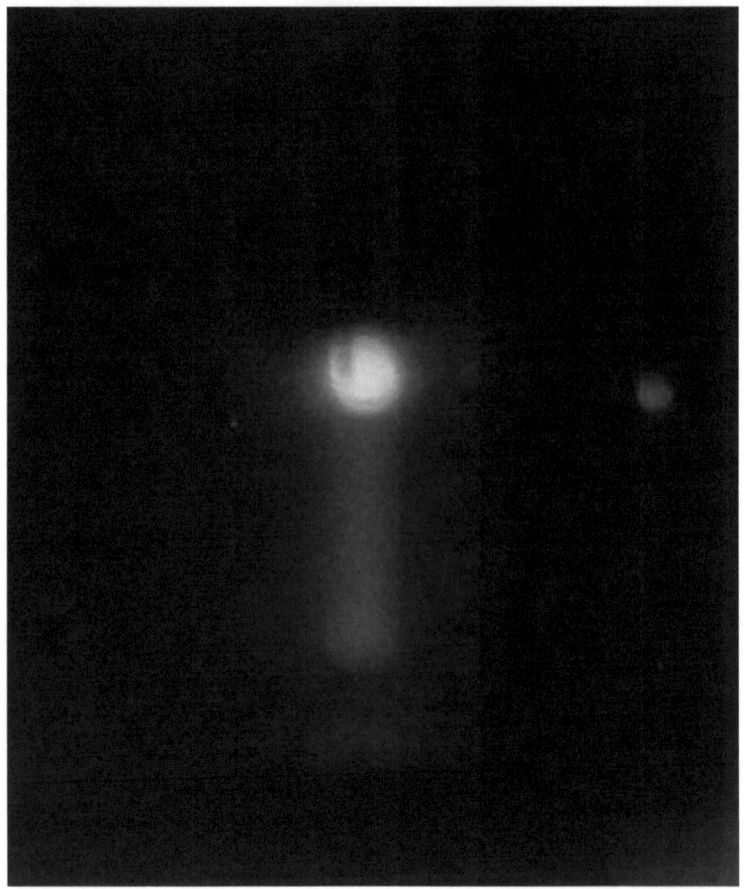

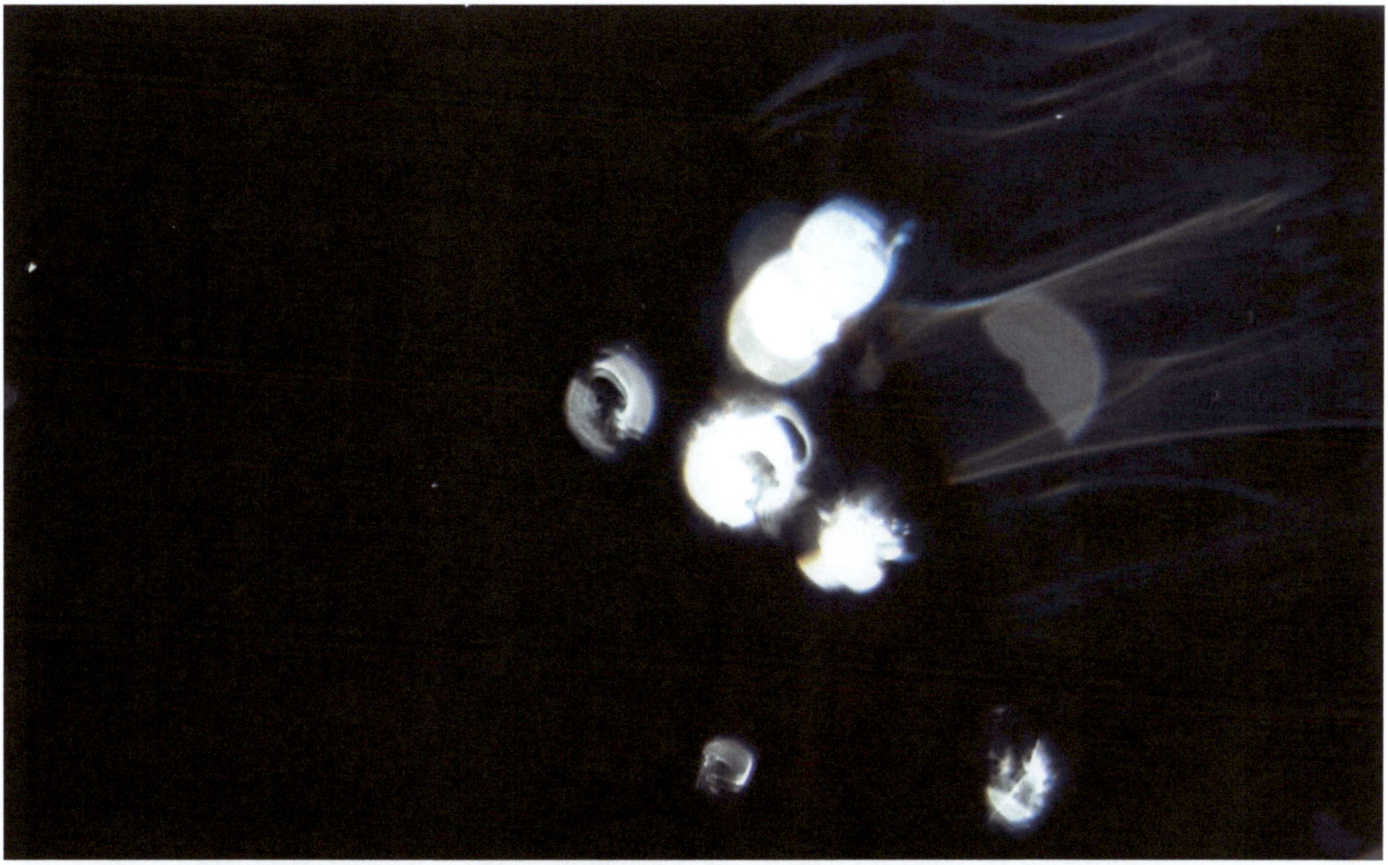

Margie Kay

Photo #3 above.   Photo #4 below– some objects appear to be moving up.

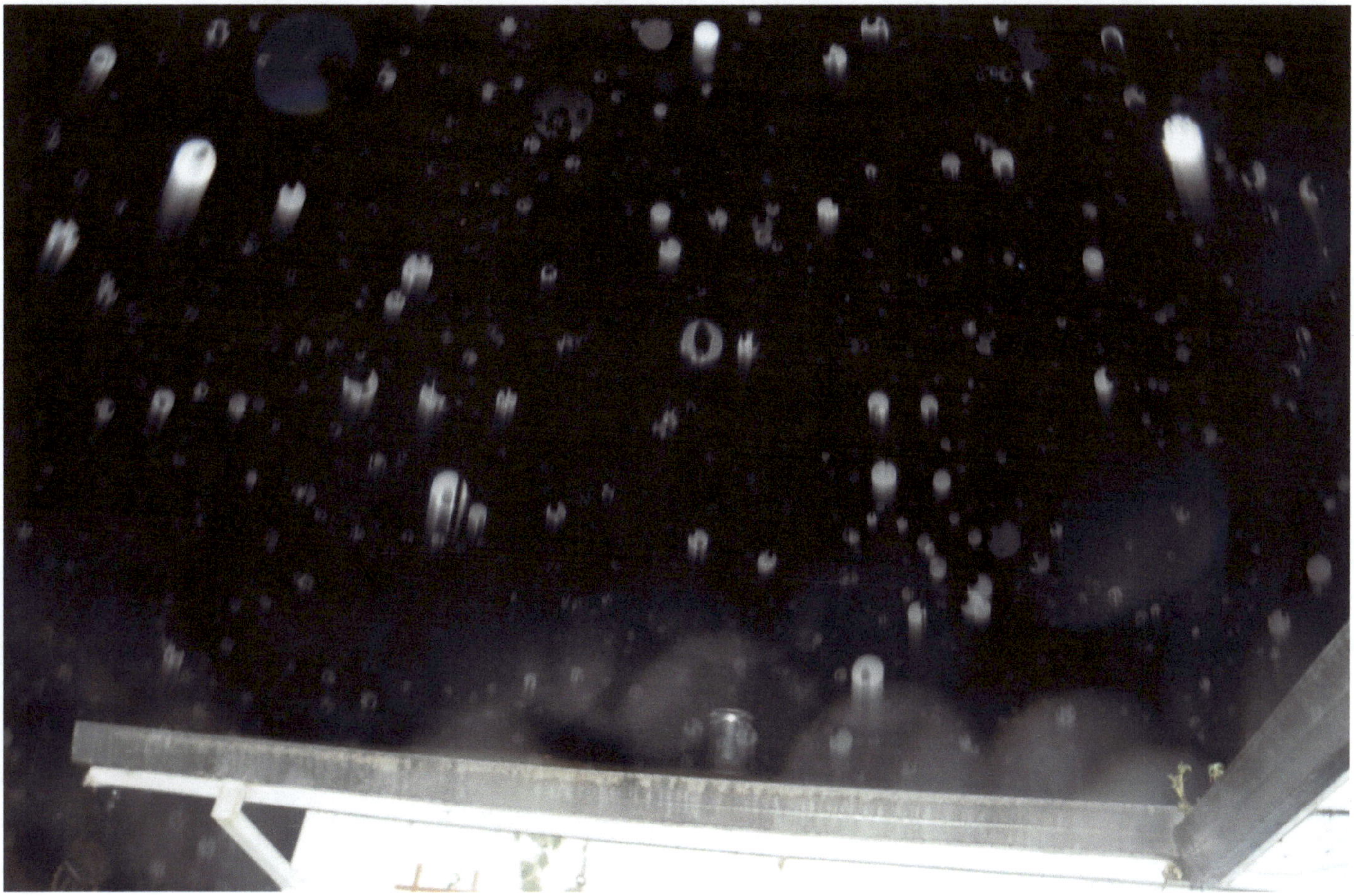

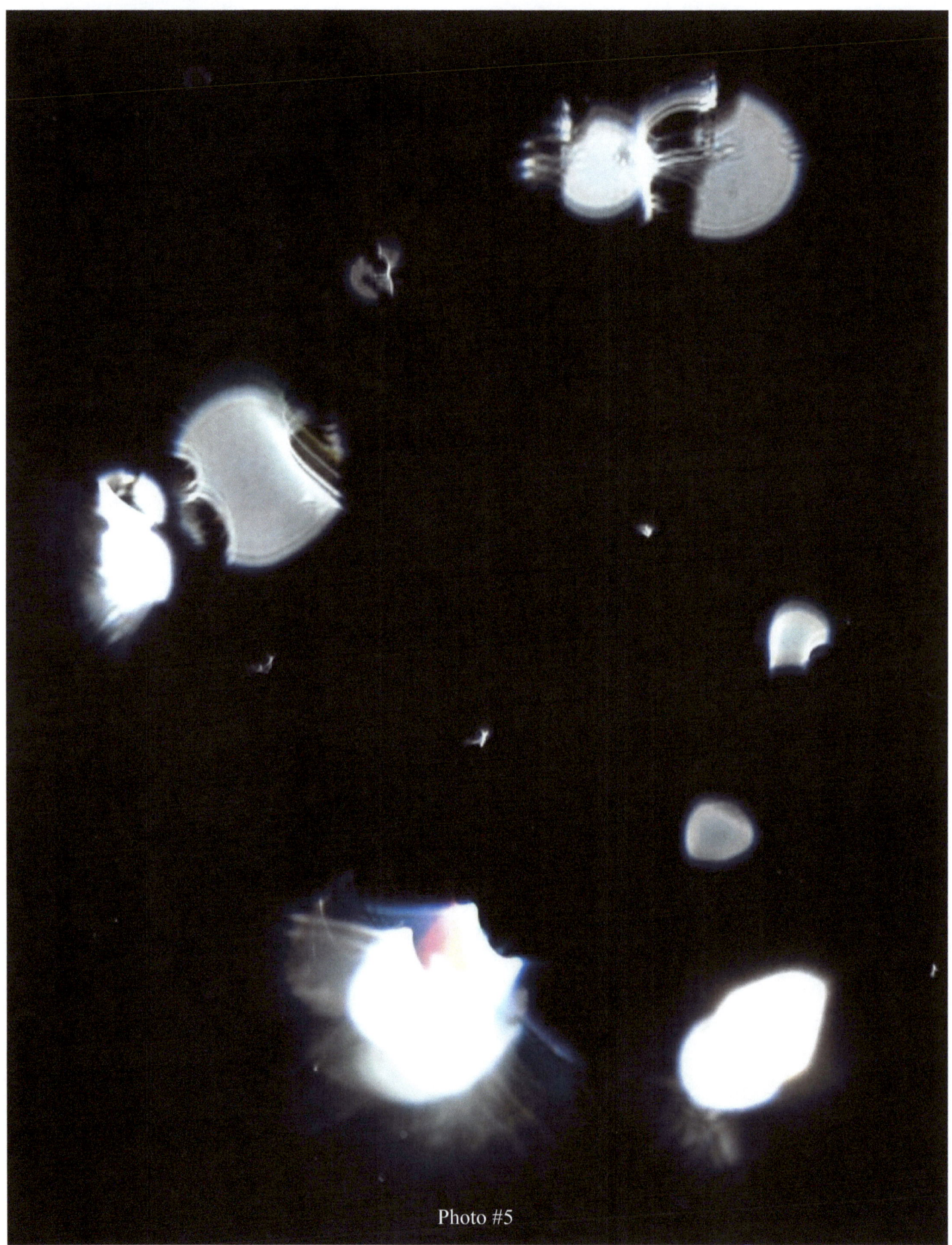

Photo #5

Margie Kay

Photo #6 above, enlargement of object in lower center below:

It appears that there is a dark orb covering the lighter orb. Note the blue and yellow edges.

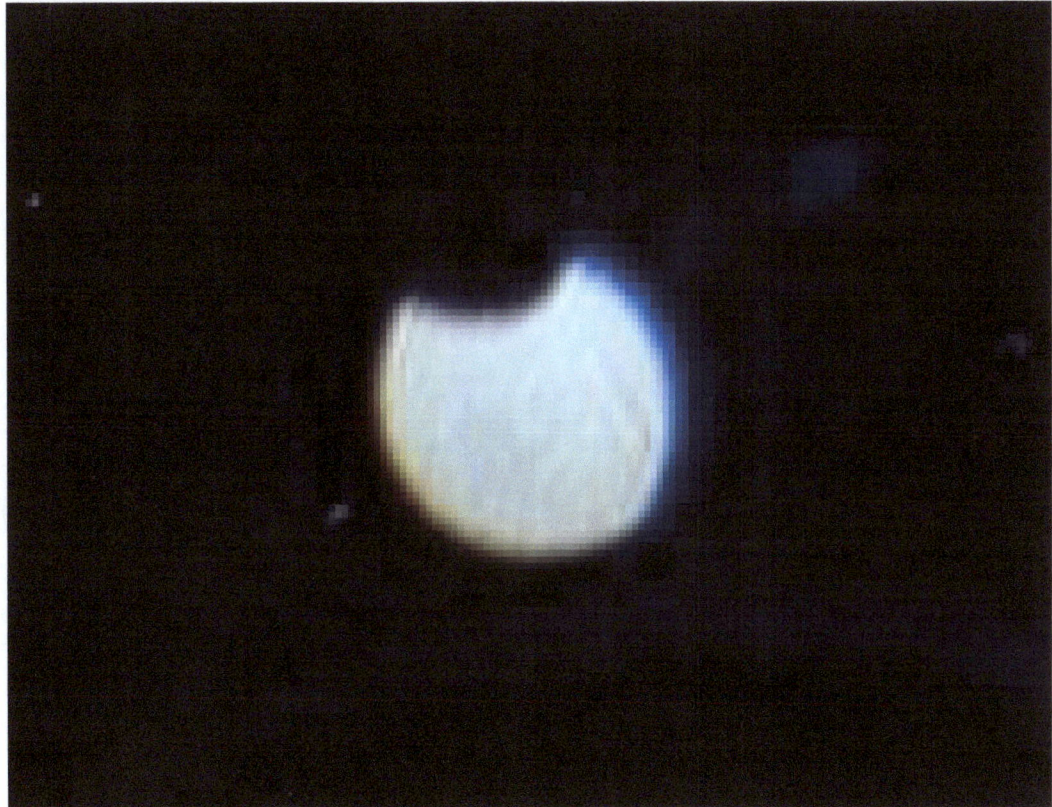

A Sonoma County Phenomenon

Photo #7

Right: close-up of two objects at the right side

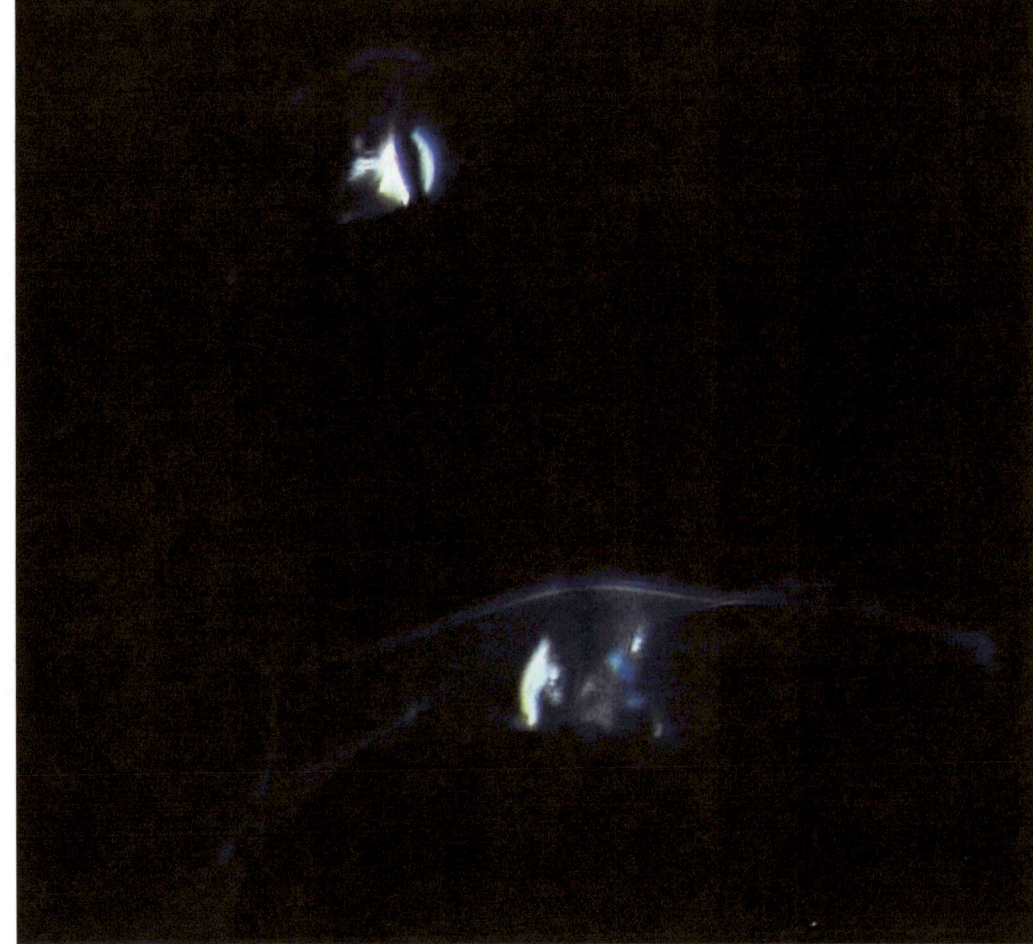

24

Photo 7: Enlargement of lower object

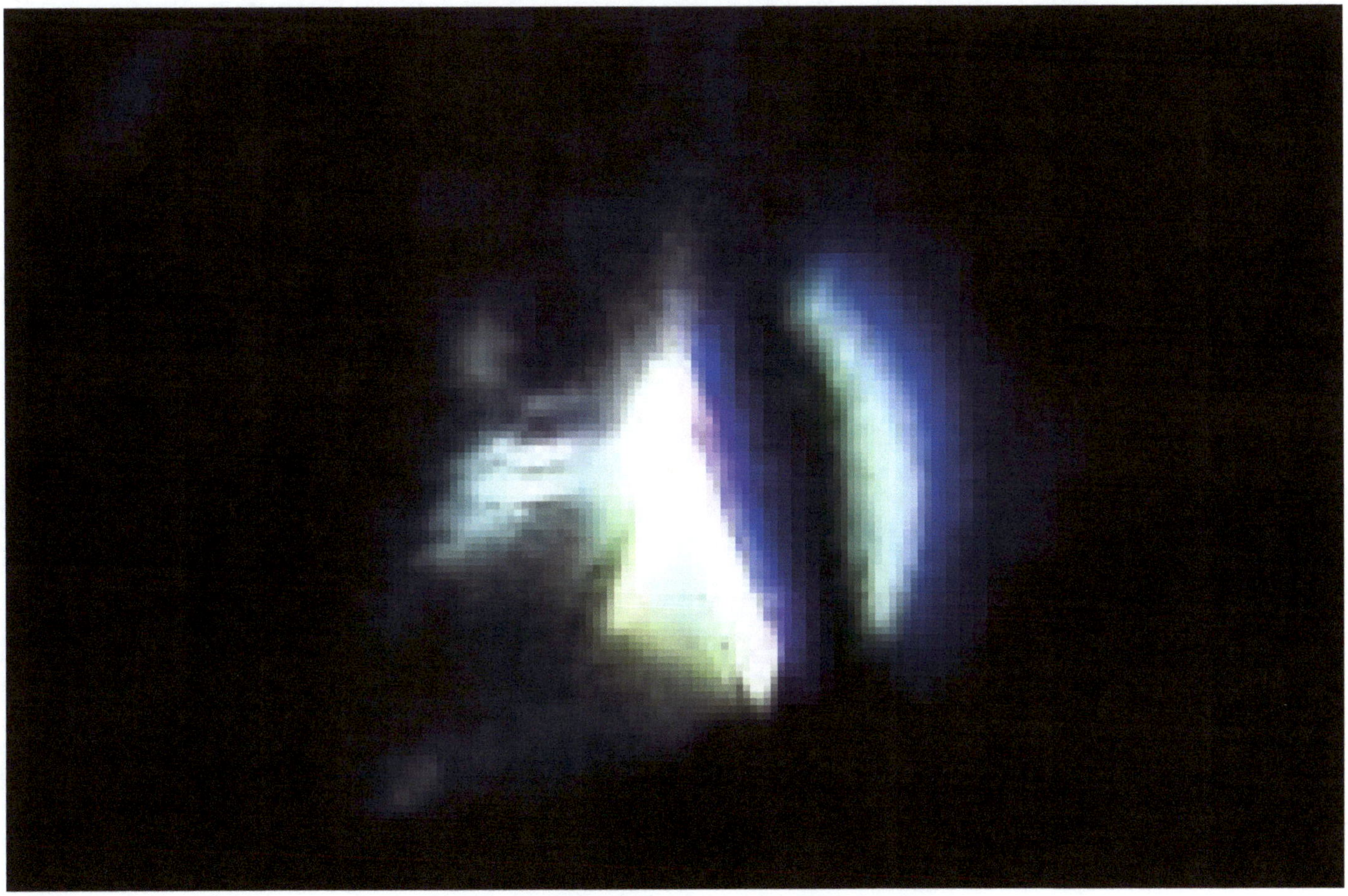

Photo 7: Enlargement of upper object

A Sonoma County Phenomenon

Photo #8

Right: enlargement of bright object in photo above.

Some things make you think *"What the Heck is that?"*

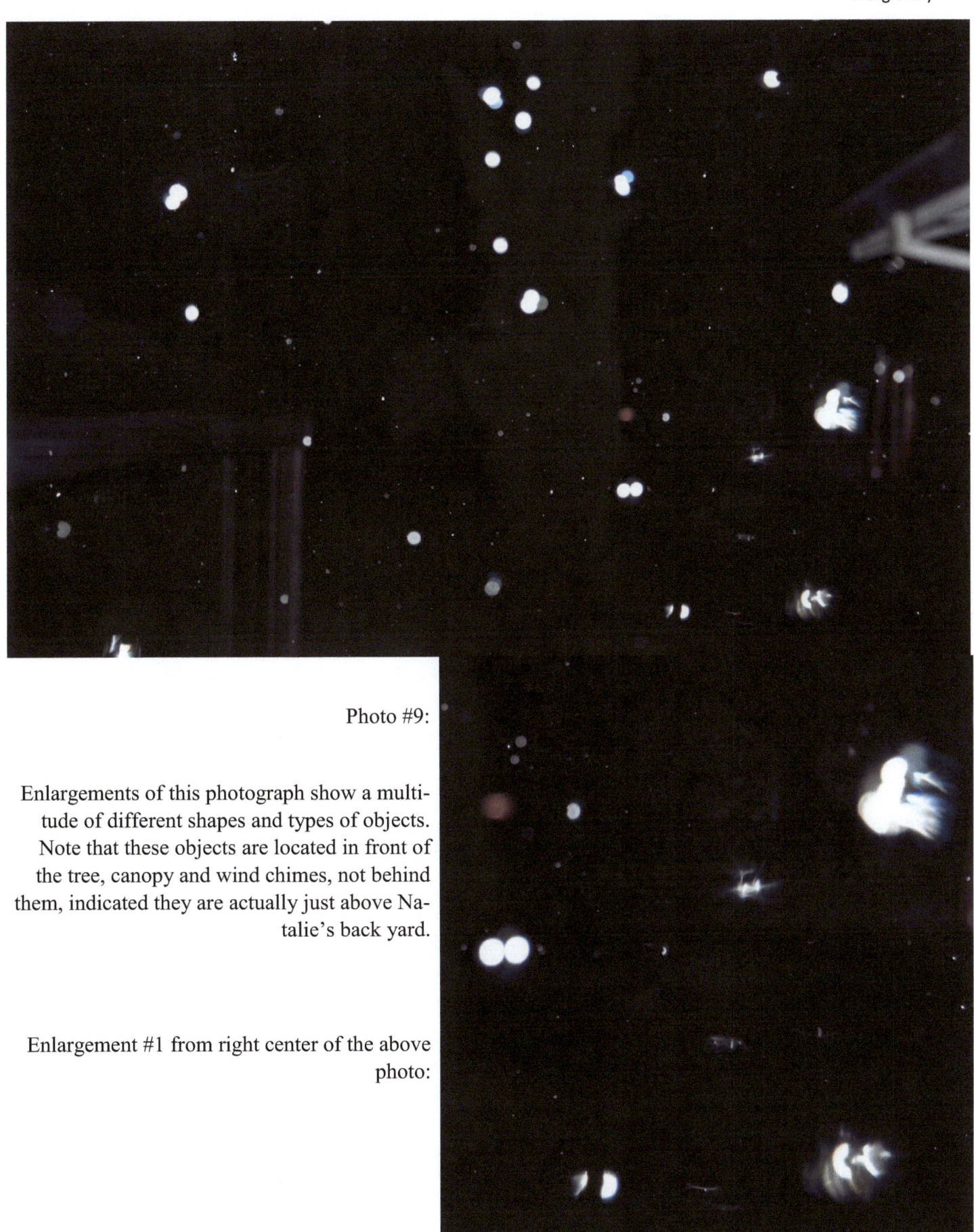

Photo #9:

Enlargements of this photograph show a multitude of different shapes and types of objects. Note that these objects are located in front of the tree, canopy and wind chimes, not behind them, indicated they are actually just above Natalie's back yard.

Enlargement #1 from right center of the above photo:

Photo #10

Right: Close up of bright object in upper left corner. It appears to be moving upward and to the right. Other objects appear to be moving up and to the left. If this were caused by camera shake, all of the objects would be moving in the same direction.

:

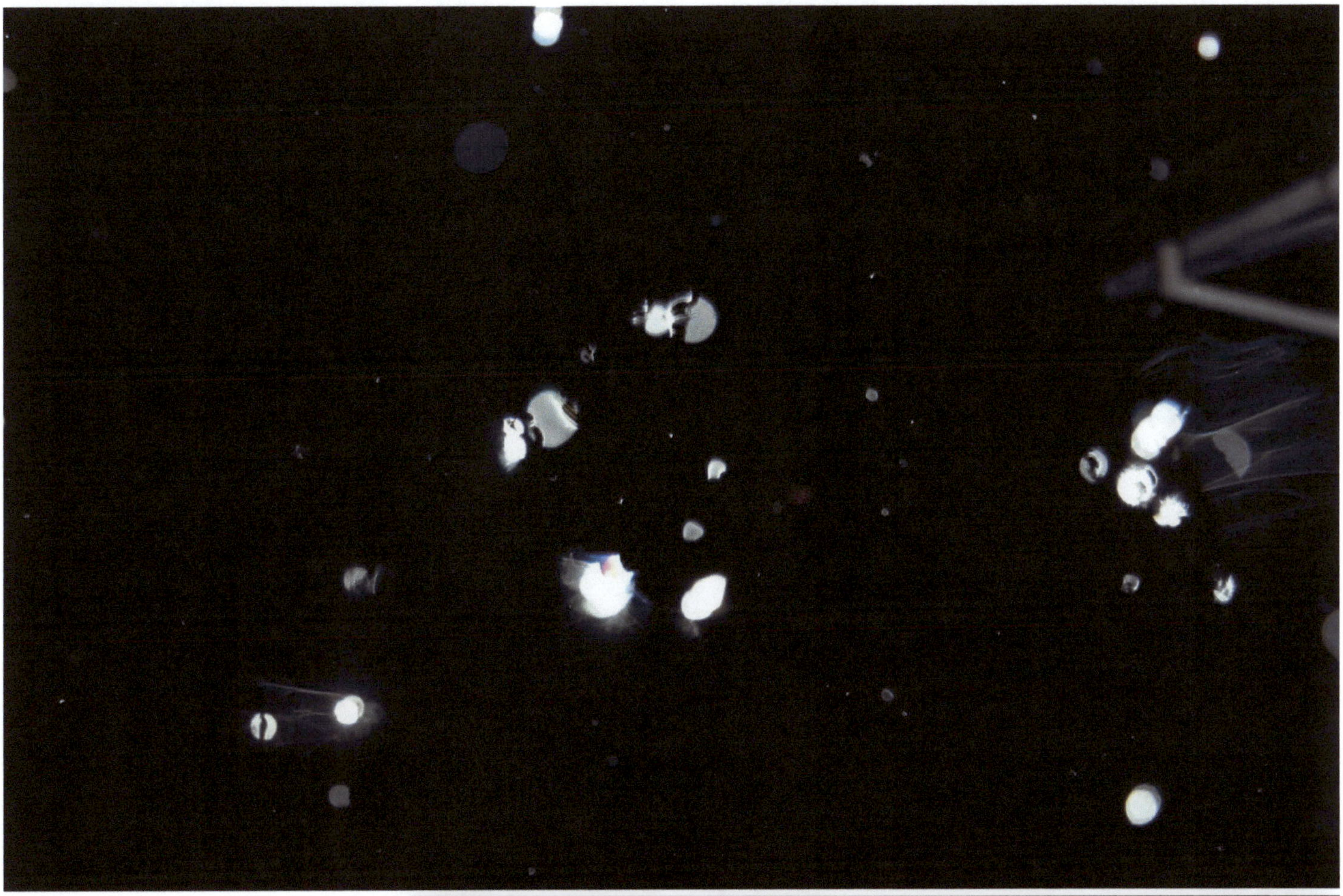

Photo #11

Right: Enlargement of lower left section.

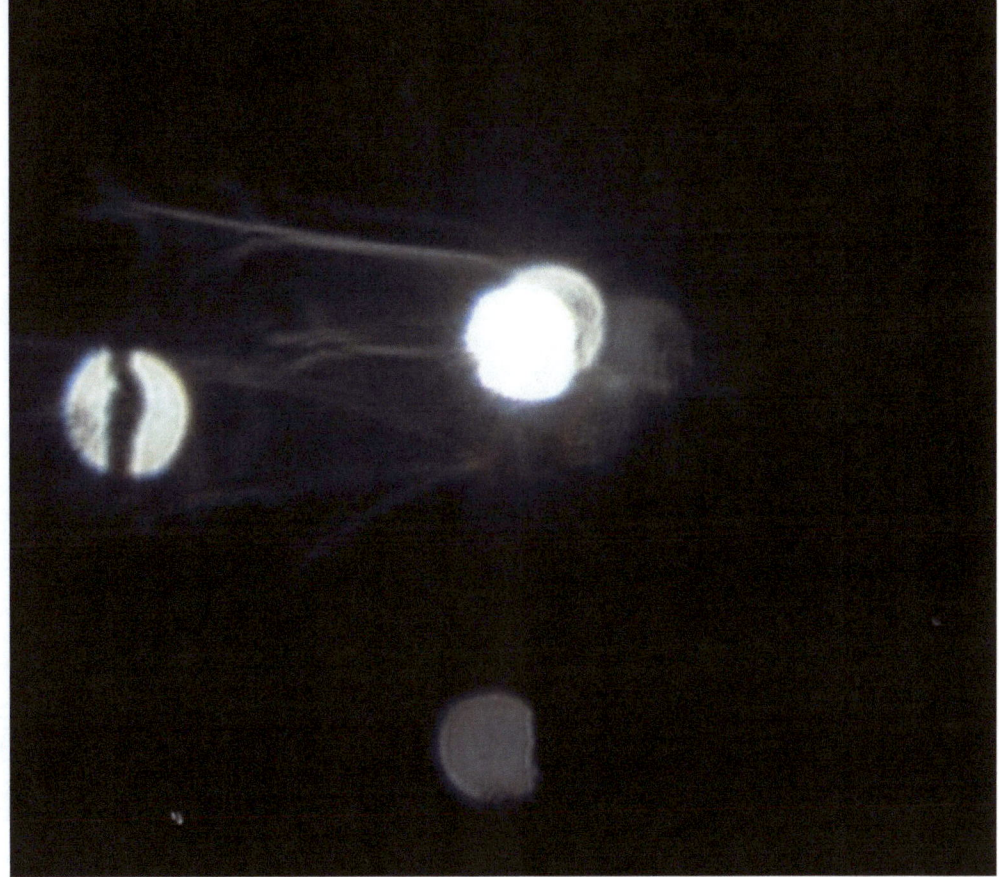

Photo #12

Below: Cantier object enlarged

Photo #13

Right: Close up of the object in the upper left of the photo above.

Note that most of the objects in the pictures are white or grey but some appear as multiple colored-objects. In this picture, it looks like there are two objects of the same size and shape, but different colors.

Photo #14

Right: Close up of the object in the center of the photo above.

This object has a fiend around it as well, similar to what the launch of the Space X rocket looked like. The aura round the rocket was explained as water crystals and exhaust. This type of other-worldly appearance is seen in many of the objects in the photos that Natalie took,.

Photo #15

Right: Close up of one object in the center of the photo above. Some of these objects almost look like microscopic organisms.

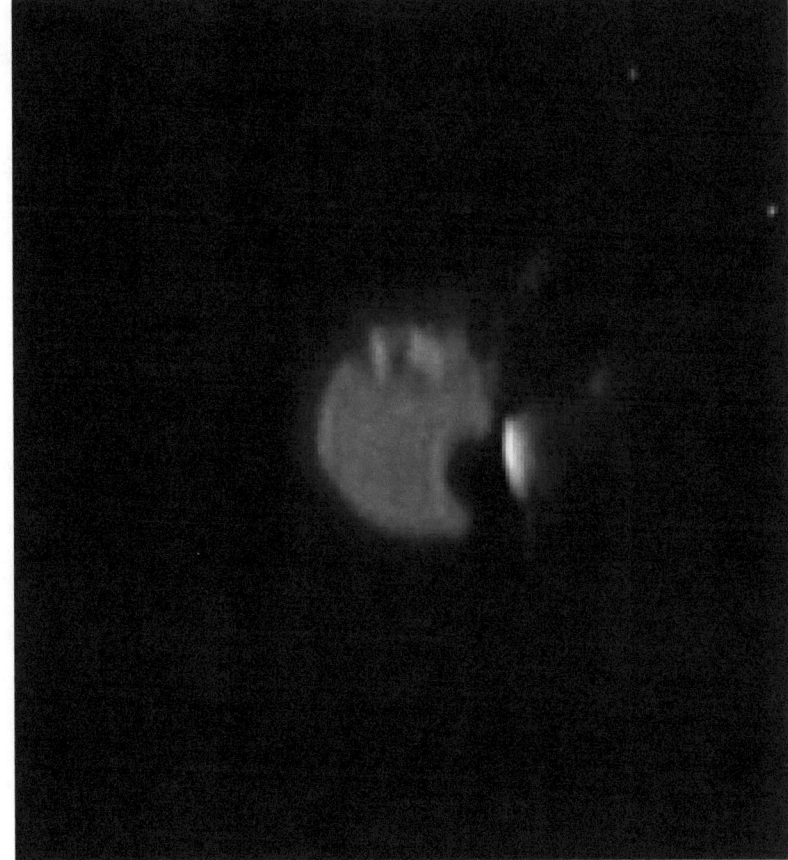

Photo #16

Right: Close up of the center objects in the photo above. Note that these objects not only appear to be moving upward, but also forward, leaving streaks of light behind them.

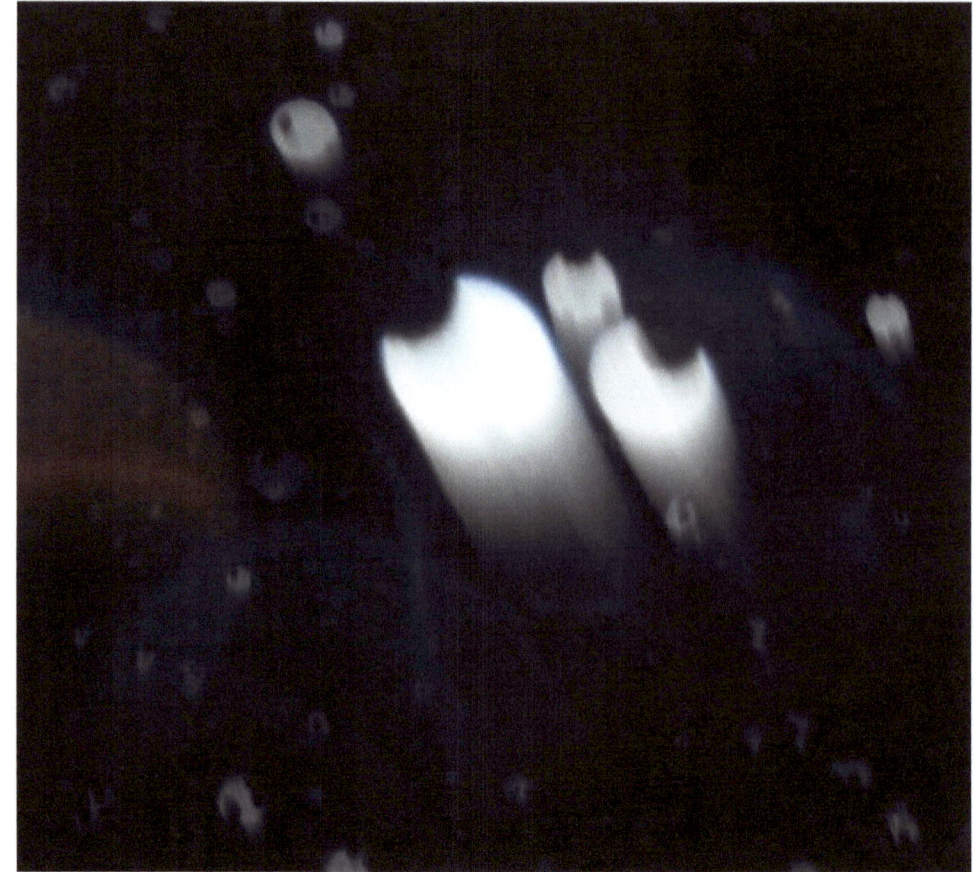

Photo #17

Right: enlargement of objects at the right side of the photo.

Note how different each of these objects are.

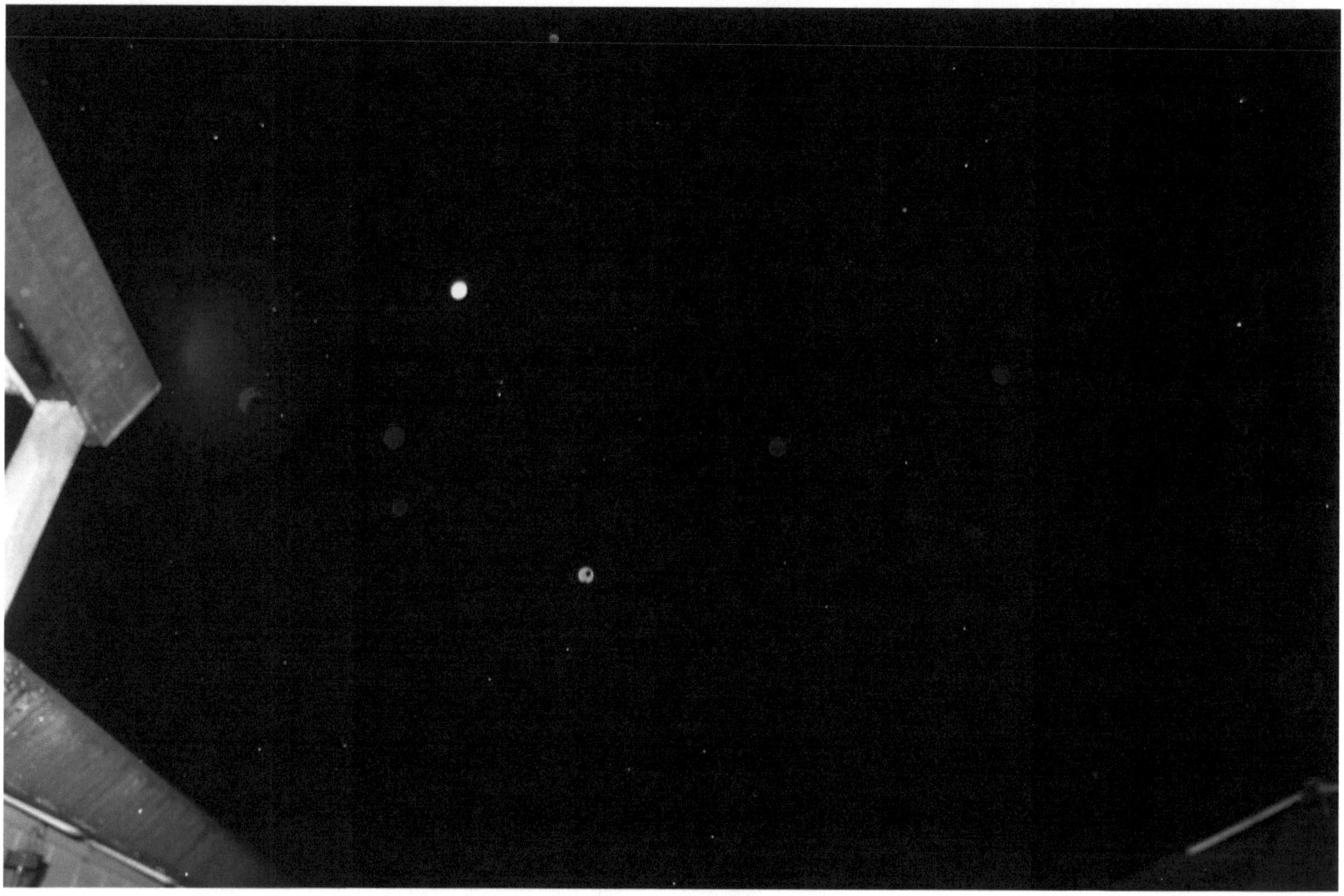

A Sonoma County Phenomenon

Photo #18

Below: objects enlarged and brightened to show more detail. It appears that there is a light object in the background and smaller objects in front of it in both instances.

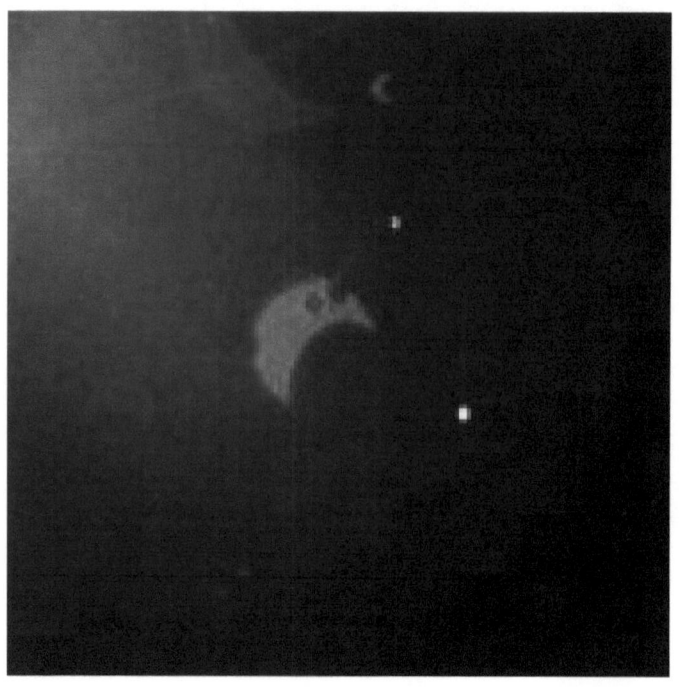
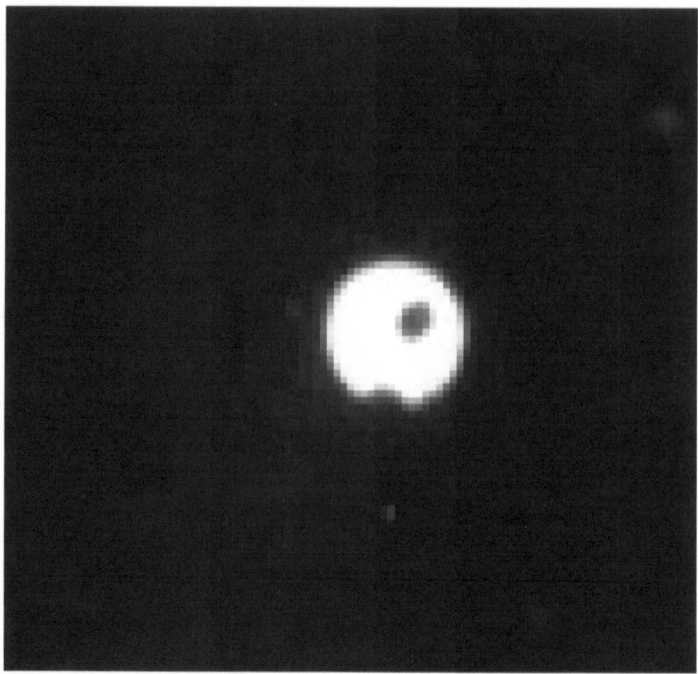

Photo #19 (above) Enlargement below

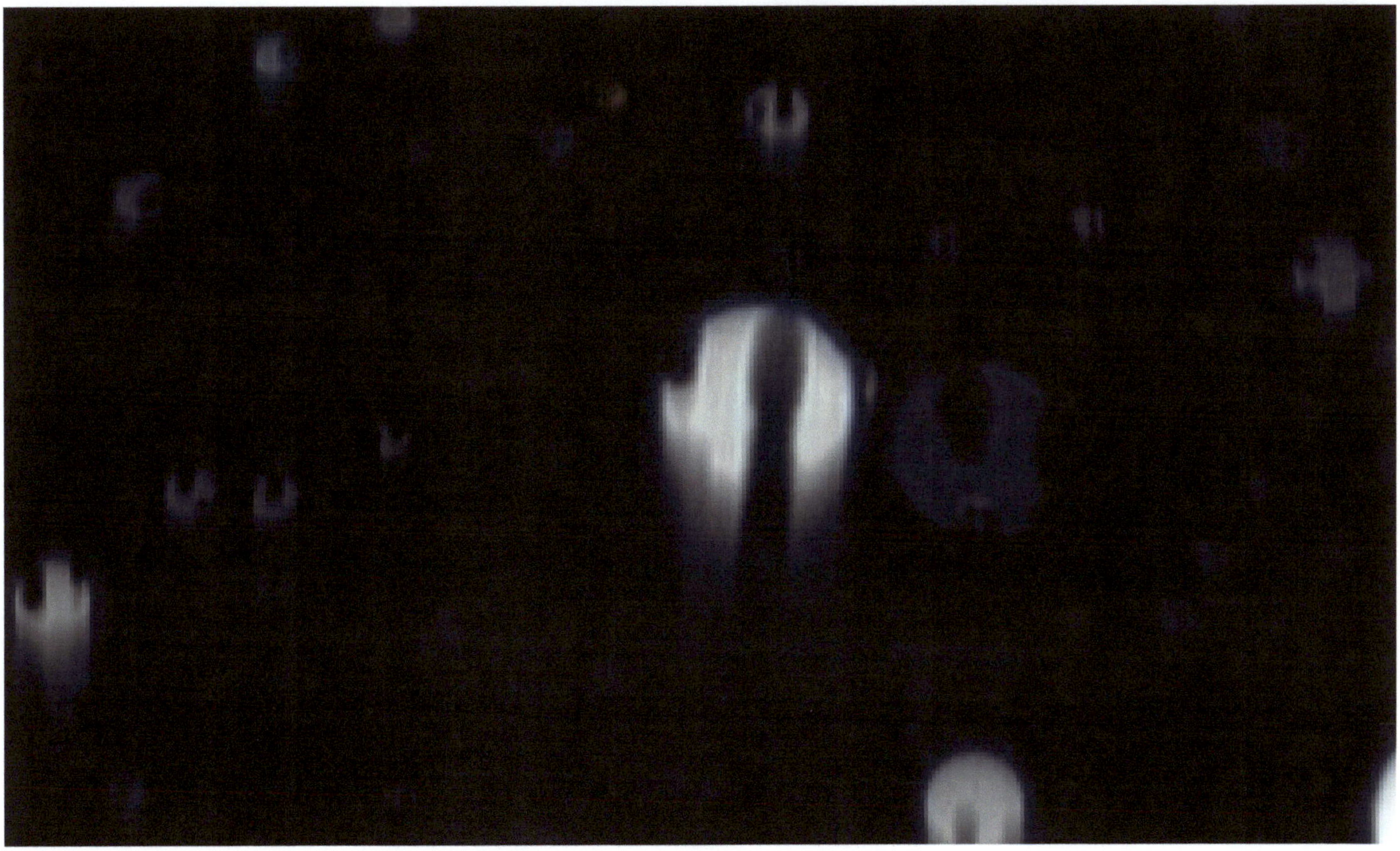

A Sonoma County Phenomenon

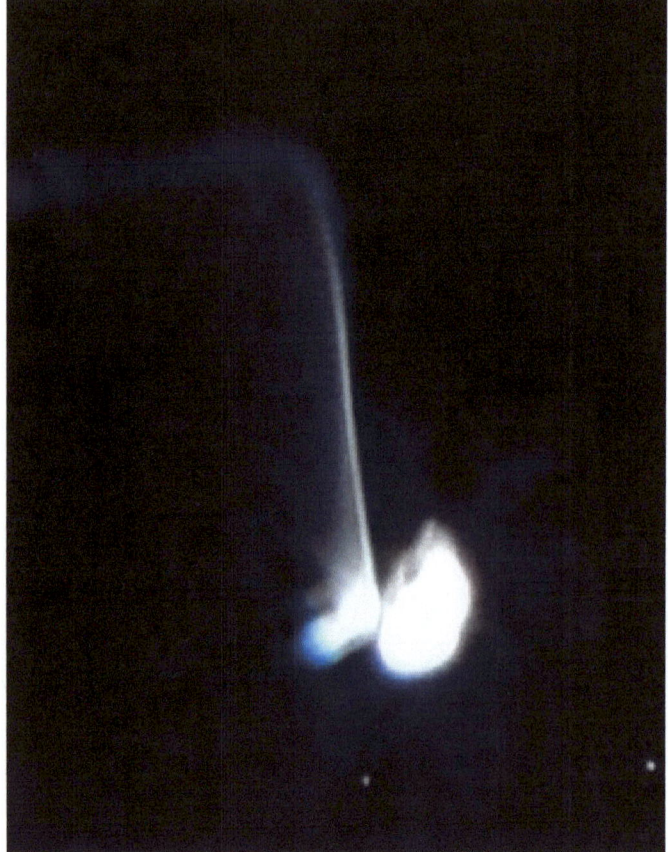

Photo #20

Left: enlargement of object at center left of the photo above.

Beautiful and mysterious!

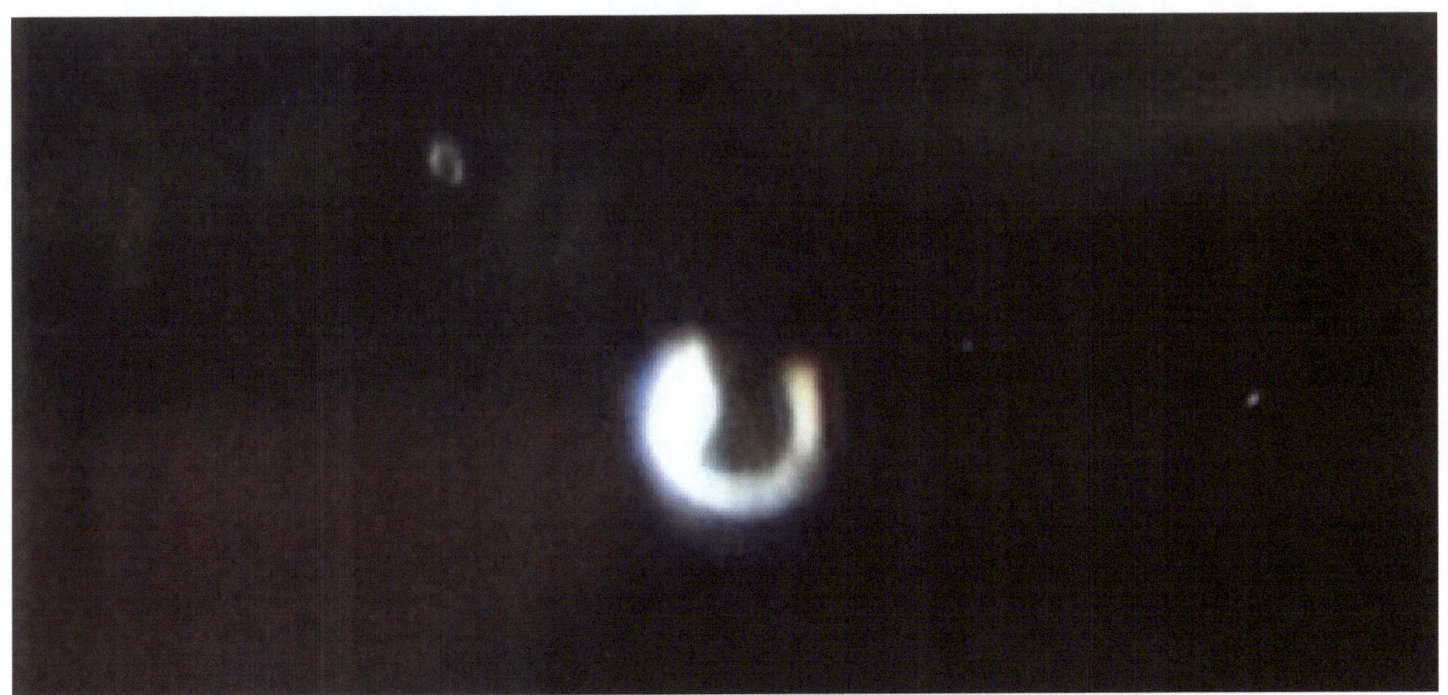

Photo 20:
Above: enlargement from lower center right. This type of photo is often captured of moving lights on UFOs.
Below: enlargement from lower left

There is almost always much more fascinating detail to these objects when they are enlarged.

Photo #21

Below: Enlargement of upper left corner of photo. Note that the object on the lower left corner seems to be emitting some type of energy field around it.

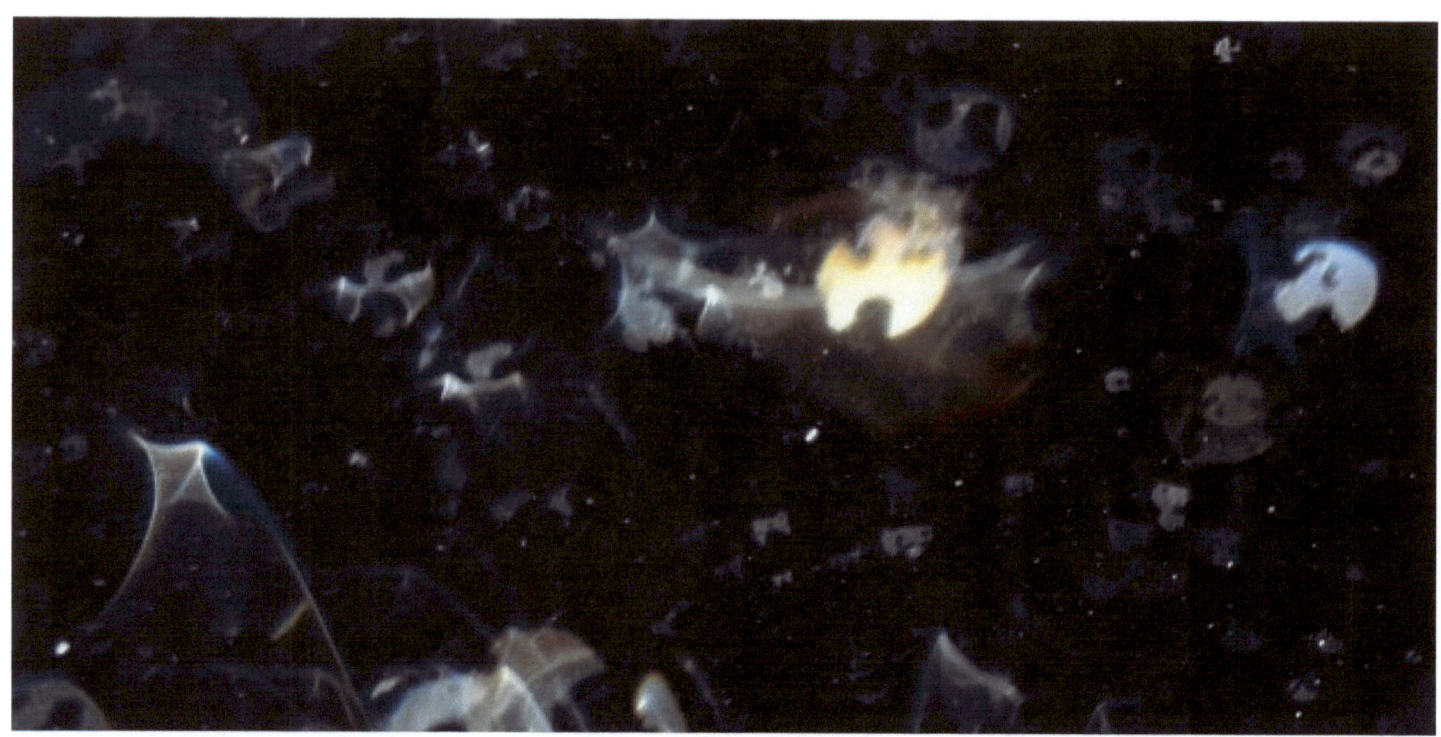

Photo #21

Enlargements of the upper right (above) and lower center (below)

Its almost as if this site is an inter-dimensional highway with hundreds or thousands of objects or beings moving through at the same time

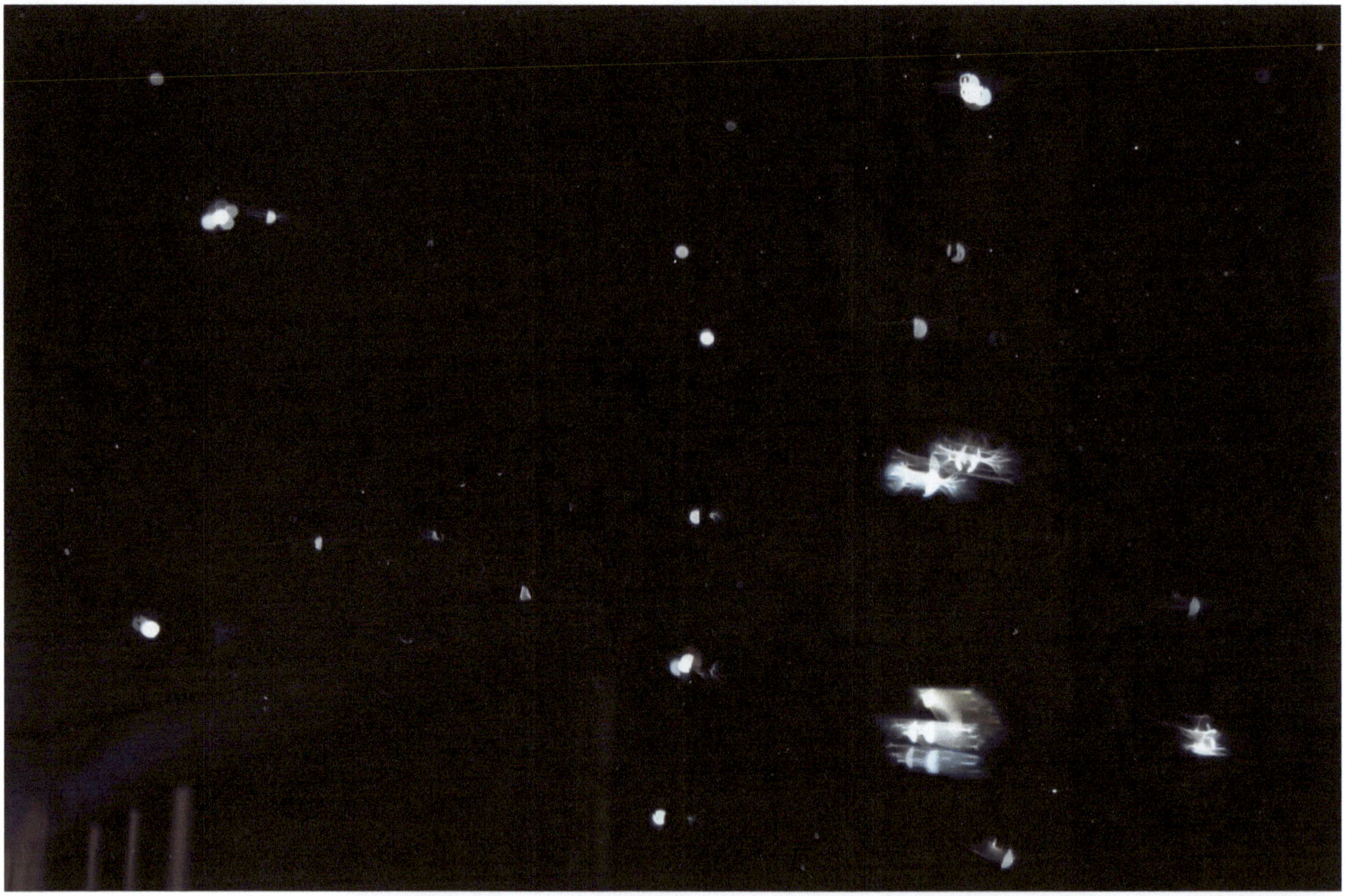

A Sonoma County Phenomenon

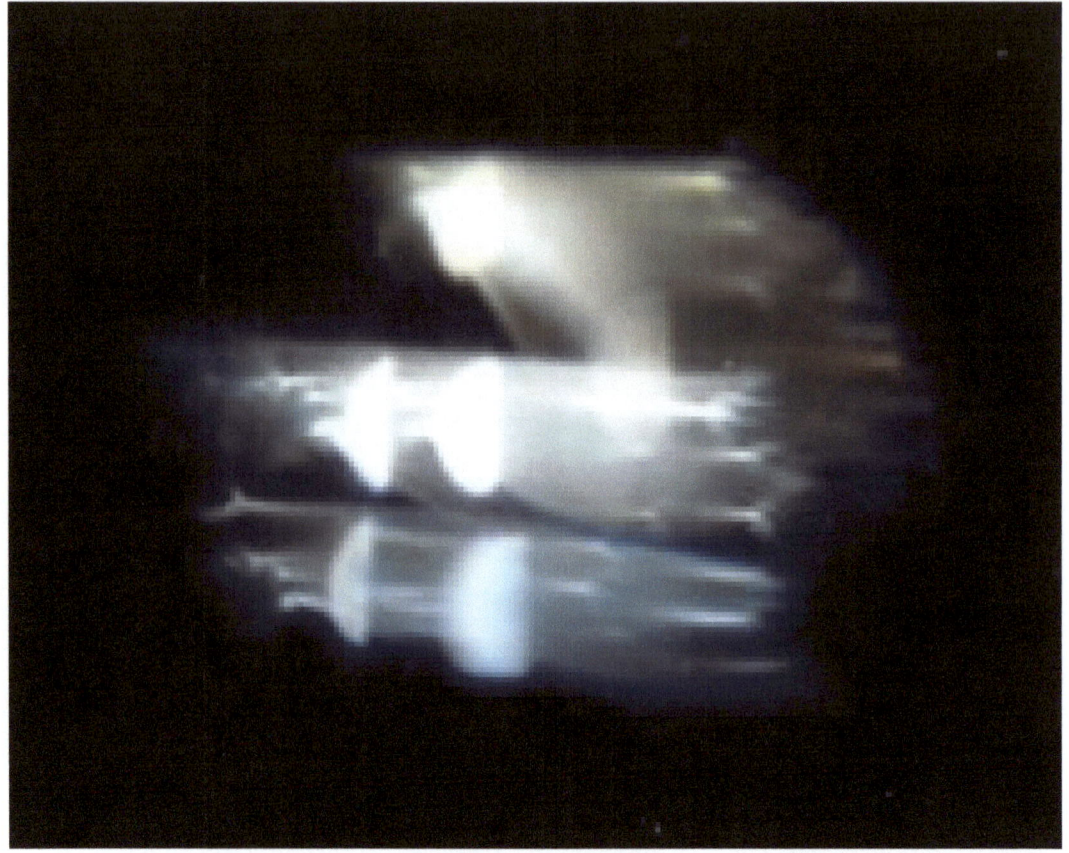

Photo #22

Right:
Enlargement of lower right section

Could these be craft of some type?

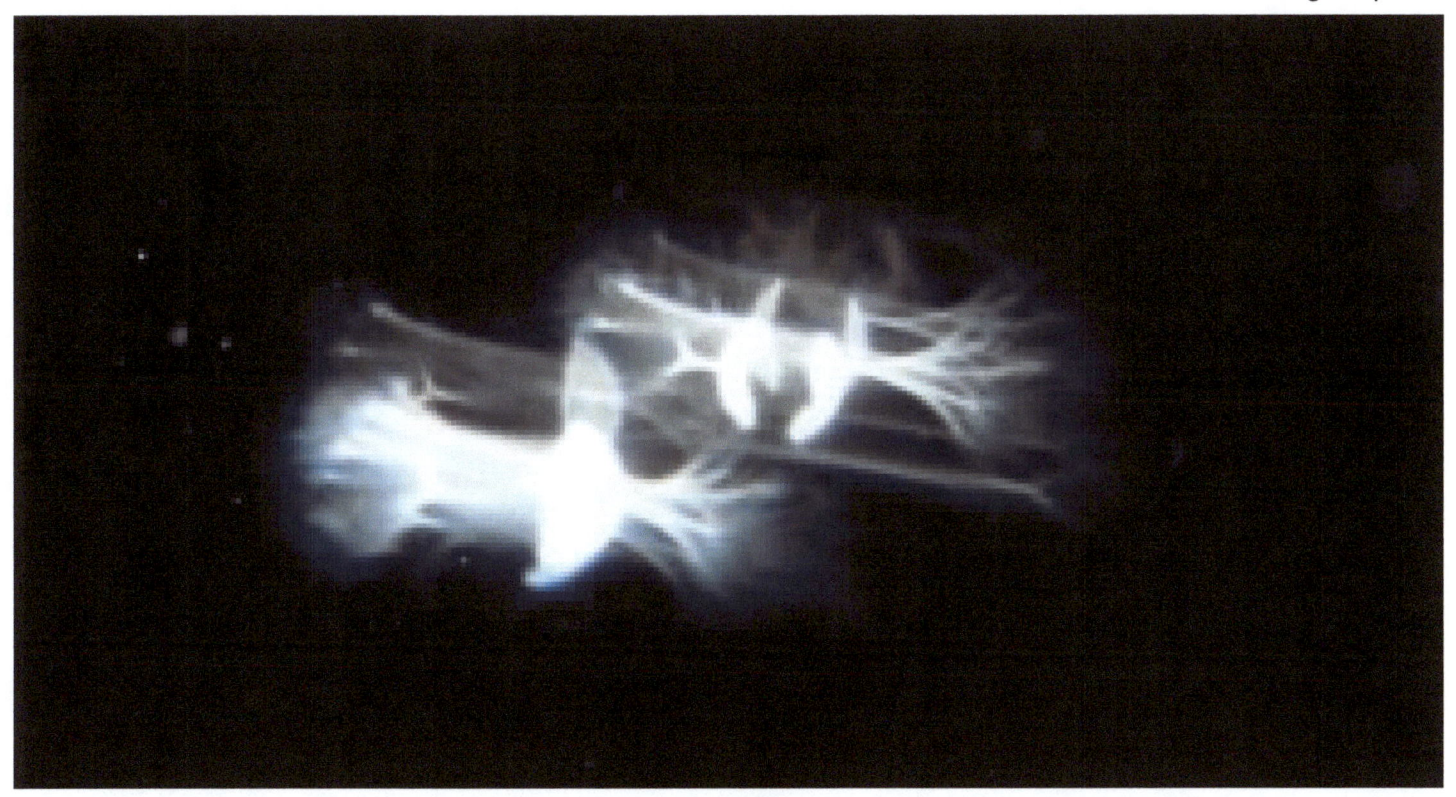

Photo #22

Above: Enlargement of upper right section
Below: Enlargement of left section.

Photo #22

Right: Enlargement of upper right section.

Photo #22: Enlargements

A Sonoma County Phenomenon

Photo #23

Right: Enlargement of lower left corner objects

Photo #23

Enlargements of objects in the photo. Note the web-like structure in the upper right photo. This is seen in numerous examples.

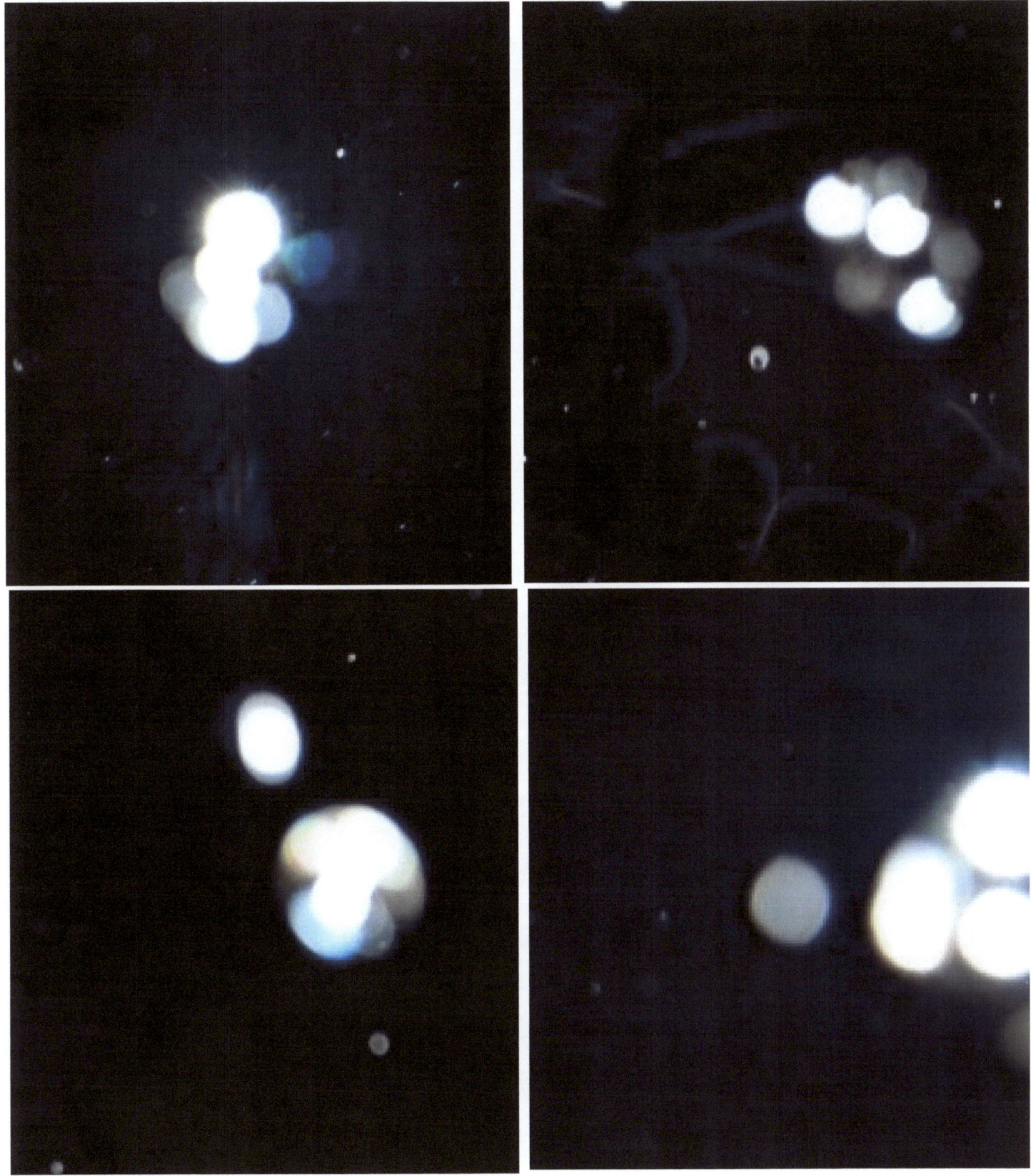

Photo #24
Below: Enlargement of center object

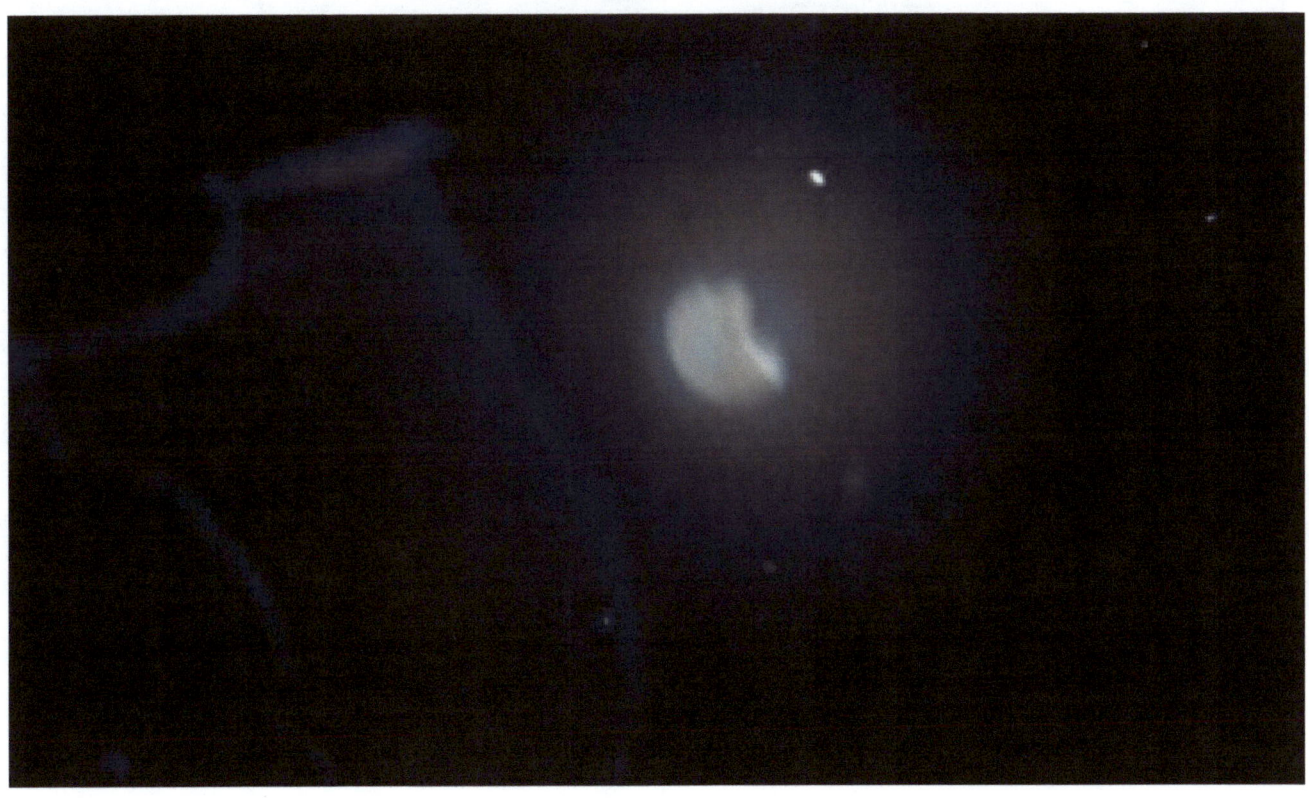

Photo #25
Below: Enlargement of center object

Photo #25
Enlargement of object up upper right corner

Photo #26
Below: Enlargement of object in upper right. Note the rainbow colors

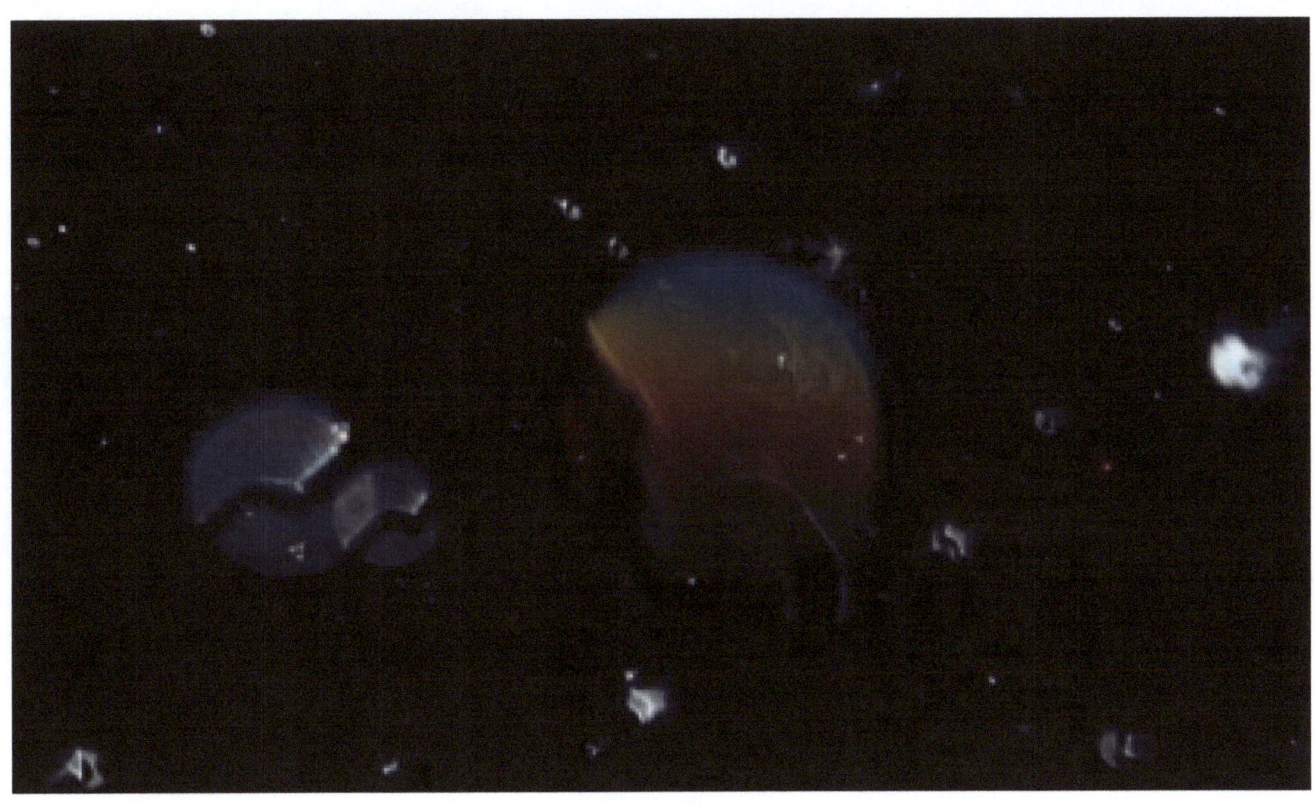

A Sonoma County Phenomenon

Photo #27
Spring of 2019
Once again, numerous objects of various sizes and shapes appear in Natalie's back yard
Right page: Enlargements of objects in this photo

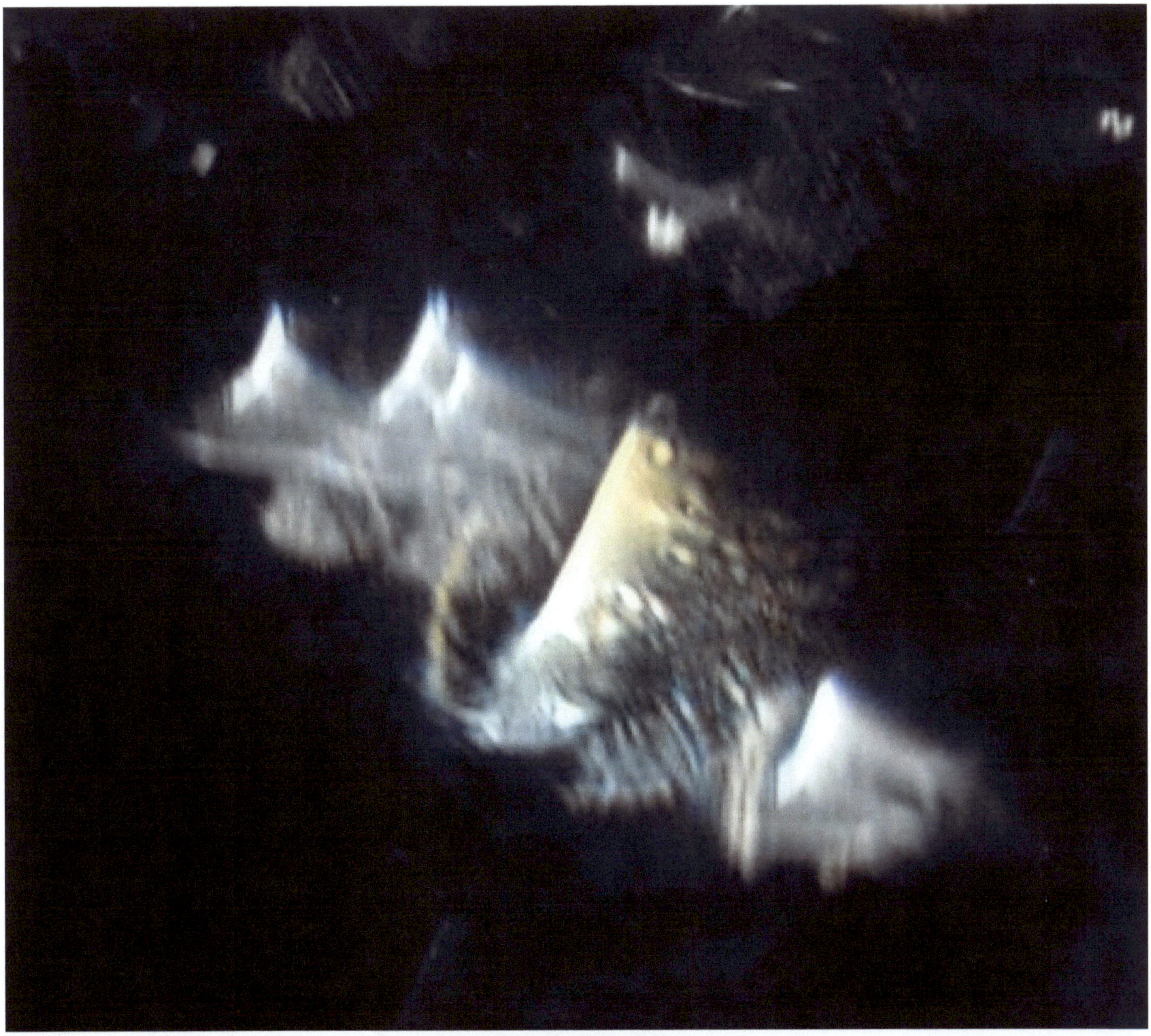

Margie Kay

Photo 27 Enlargements

Photo #28
Spring of 2019

Note the reddish fog near the stop sign and in front of the fence

(Photo brightened to show more detail)

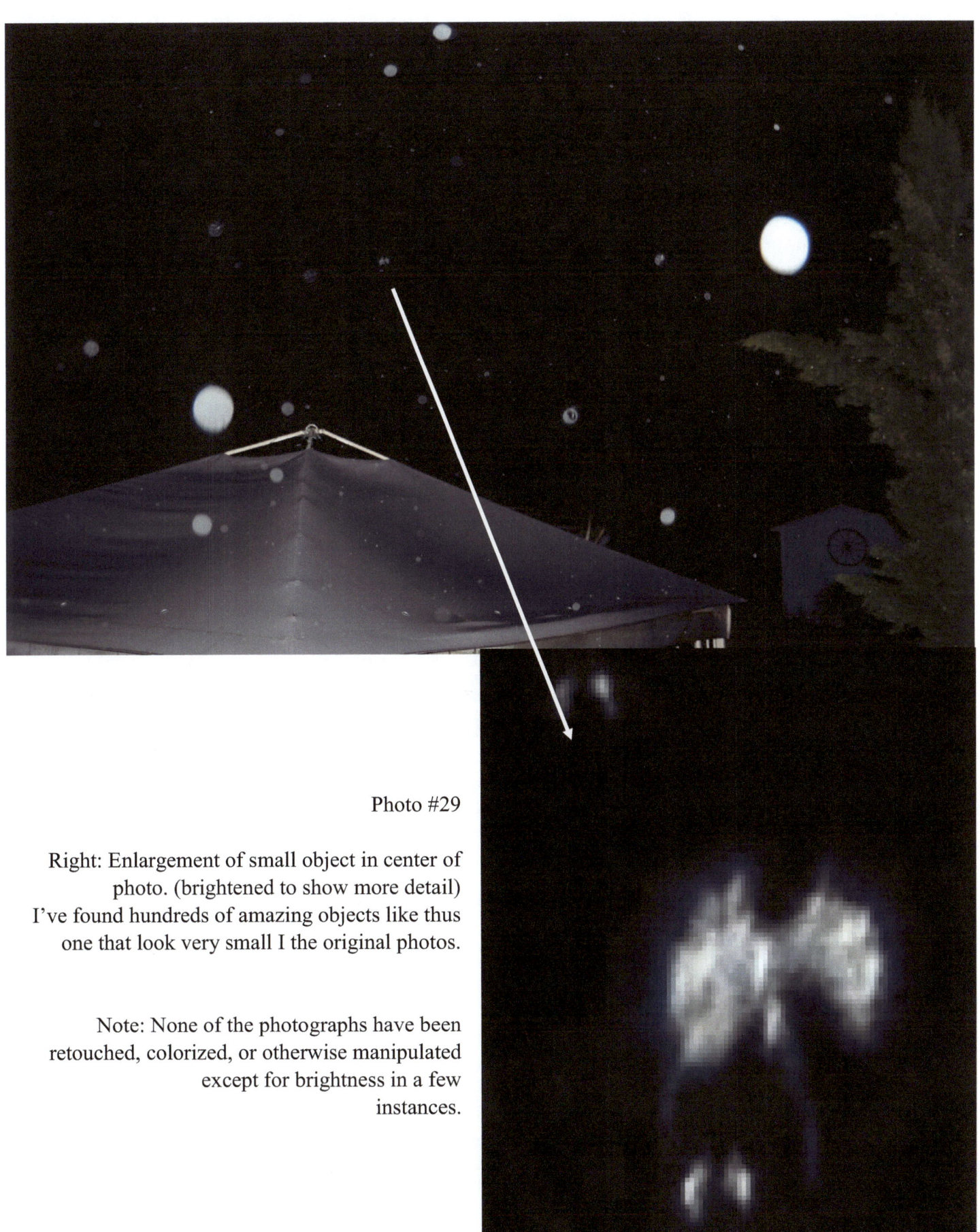

Photo #29

Right: Enlargement of small object in center of photo. (brightened to show more detail) I've found hundreds of amazing objects like thus one that look very small I the original photos.

Note: None of the photographs have been retouched, colorized, or otherwise manipulated except for brightness in a few instances.

A Sonoma County Phenomenon

Photo #30
Spring of 2019

Right: Enlargement
of bright object

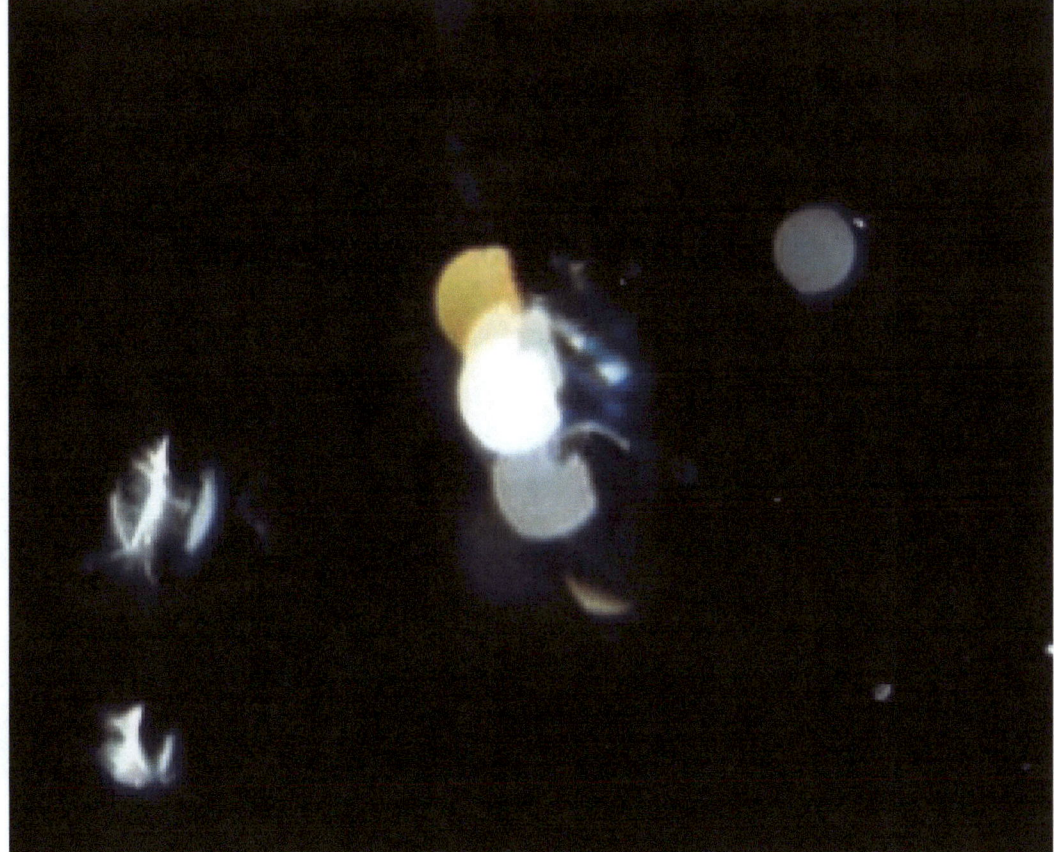

Photo #31

Right:
Enlargement of object in lower right corner

Photo #32
Above: Enlargement of object in front of house
Below: Center objects

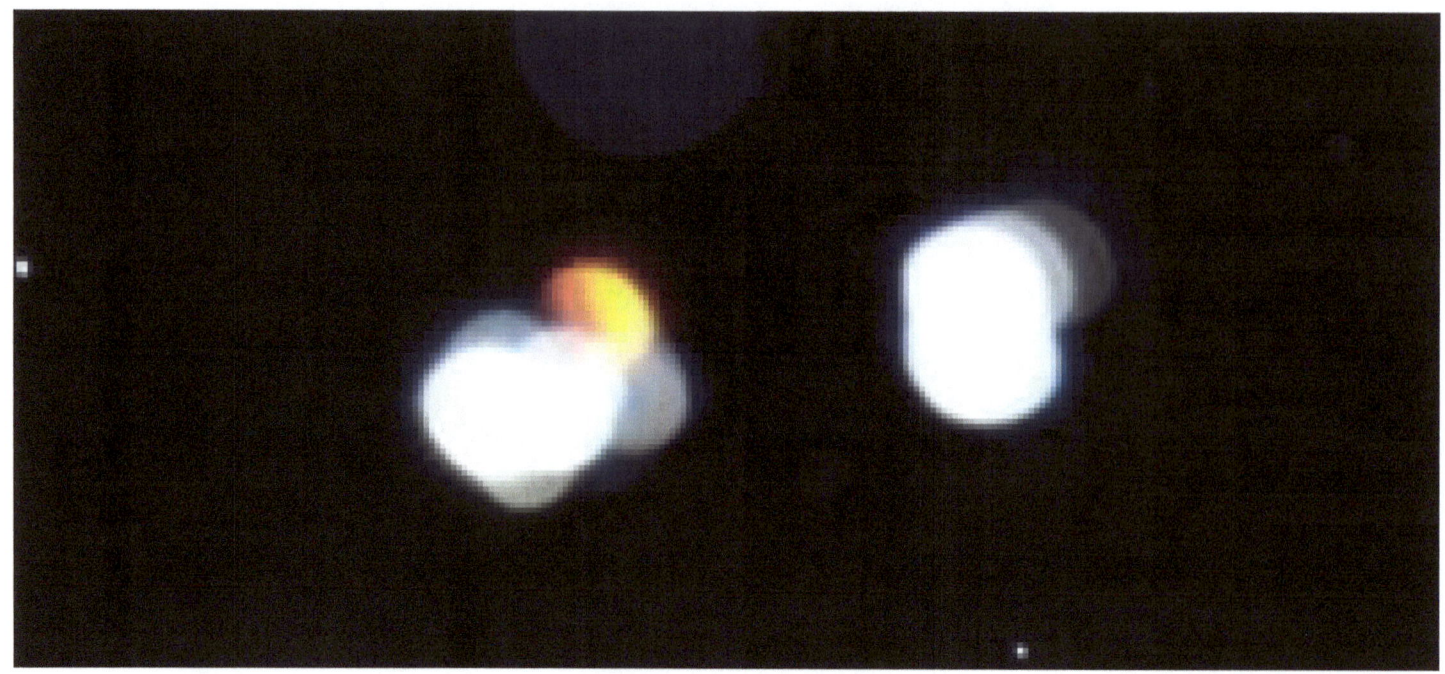

Apparitions

The following photographs contain foggy mists or apparitions. I've singled these out since they are so different from the other photos that Natalie has taken. In some instances, people see moving lights, faces or even full bodies, but most are unrecognizable. What causes these apparitions remains unexplained.

Photo 33: Apparition in front of the canopy

Photo 34: Note that this apparition is in front of the umbrella and chimes

Photo 35: The shape at the right almost resembles a gnome with a pointed hat and beard

Photo 36: Note that this apparition is in front of the umbrella and chimes

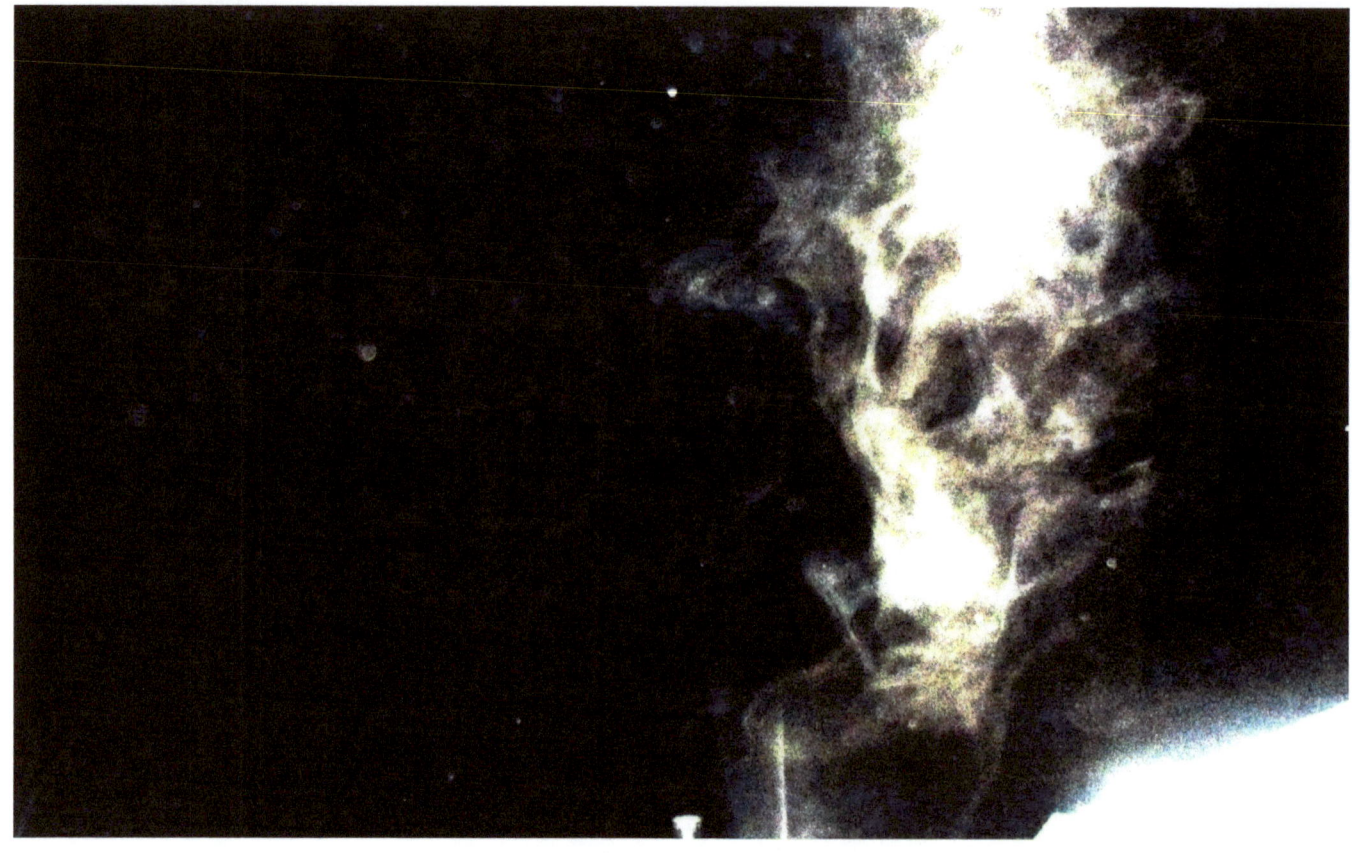

Photo 38:(top) and 39 (bottom)
Photos such as theses are sometimes captured by ghost hunters. It is believed that they may be spirits.

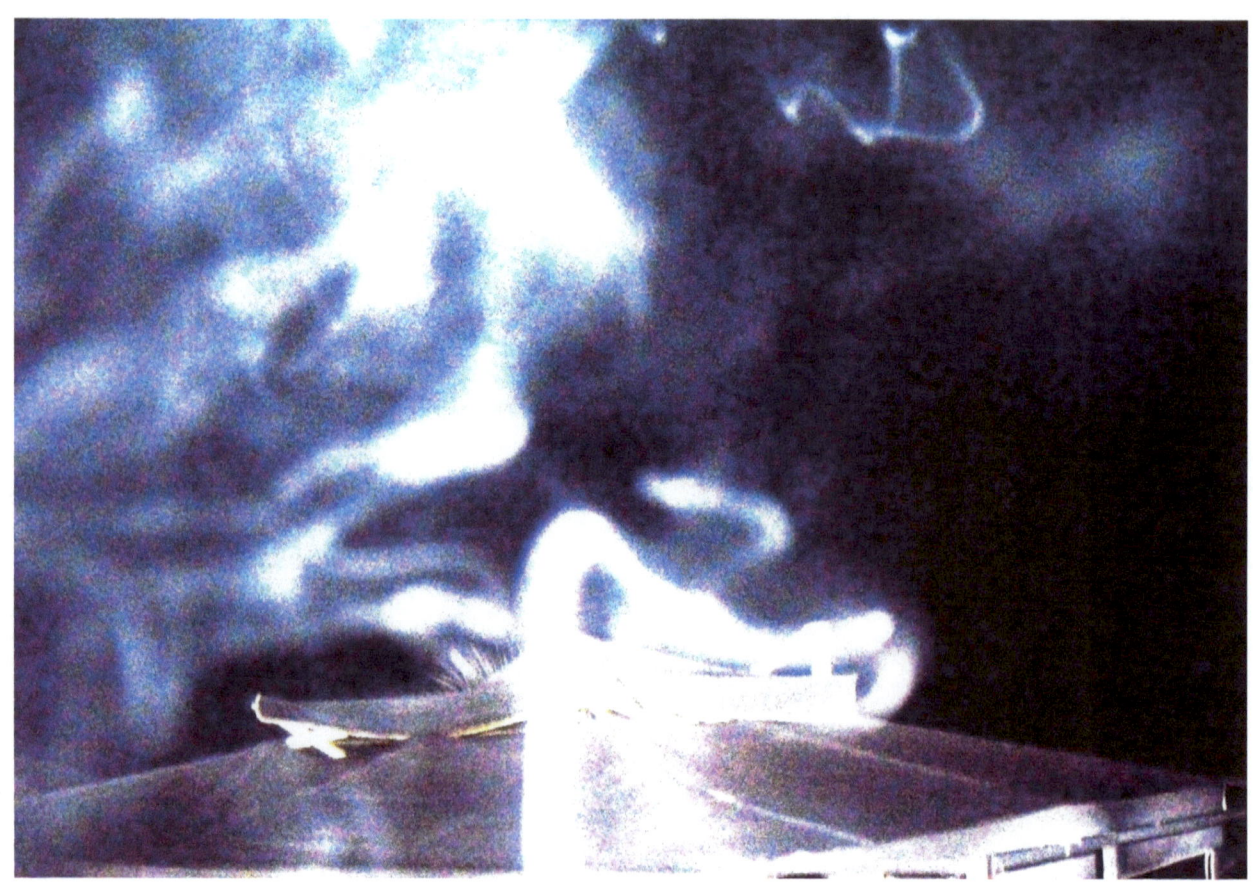

Photo 40 (right) and enlargement (left)

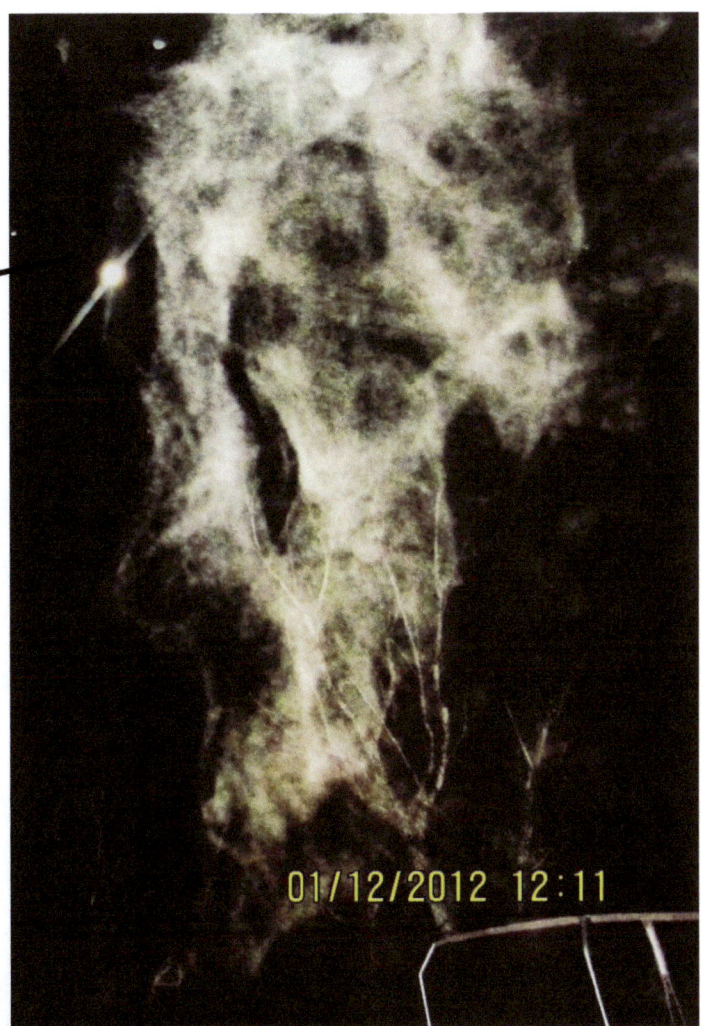
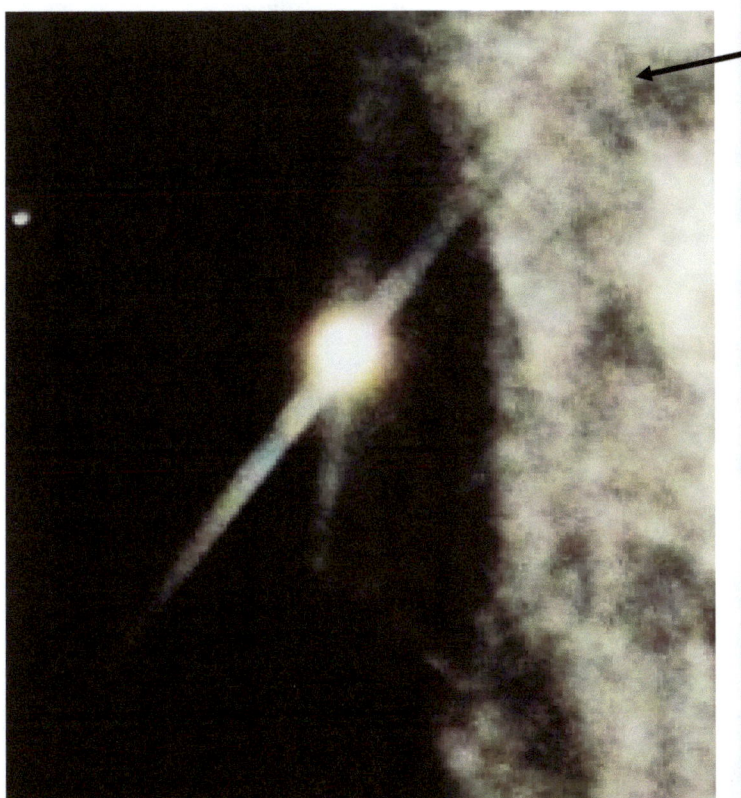

Photo 41
Foggy apparition over the fence

A Sonoma County Phenomenon

Photo 42:

Many who have seen this photo says it looks like a giant head on the left side of the picture. Note the hundreds of orbs in the rest of the picture, in fact—there are many orbs in front of the apparition as well. Note that there are no orbs at the bottom of the photo so this rules out reflections off of rain drops.

Right: Closeup of right section

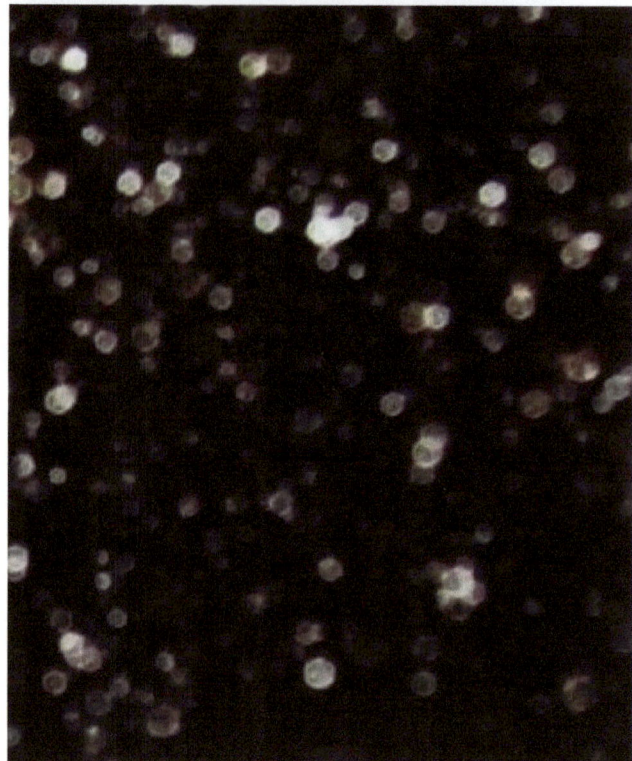

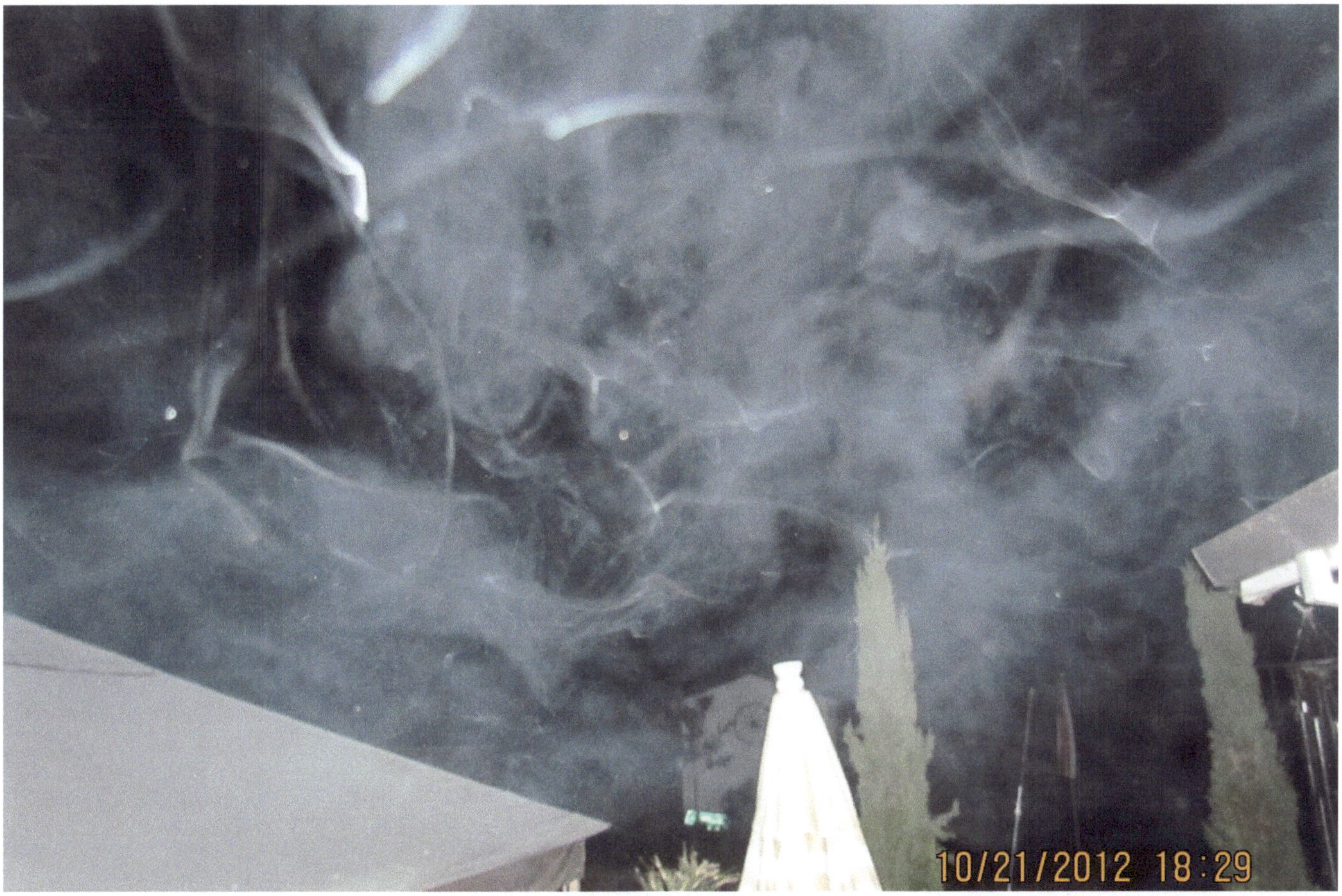

Photo 43 (top) and 44 (bottom)

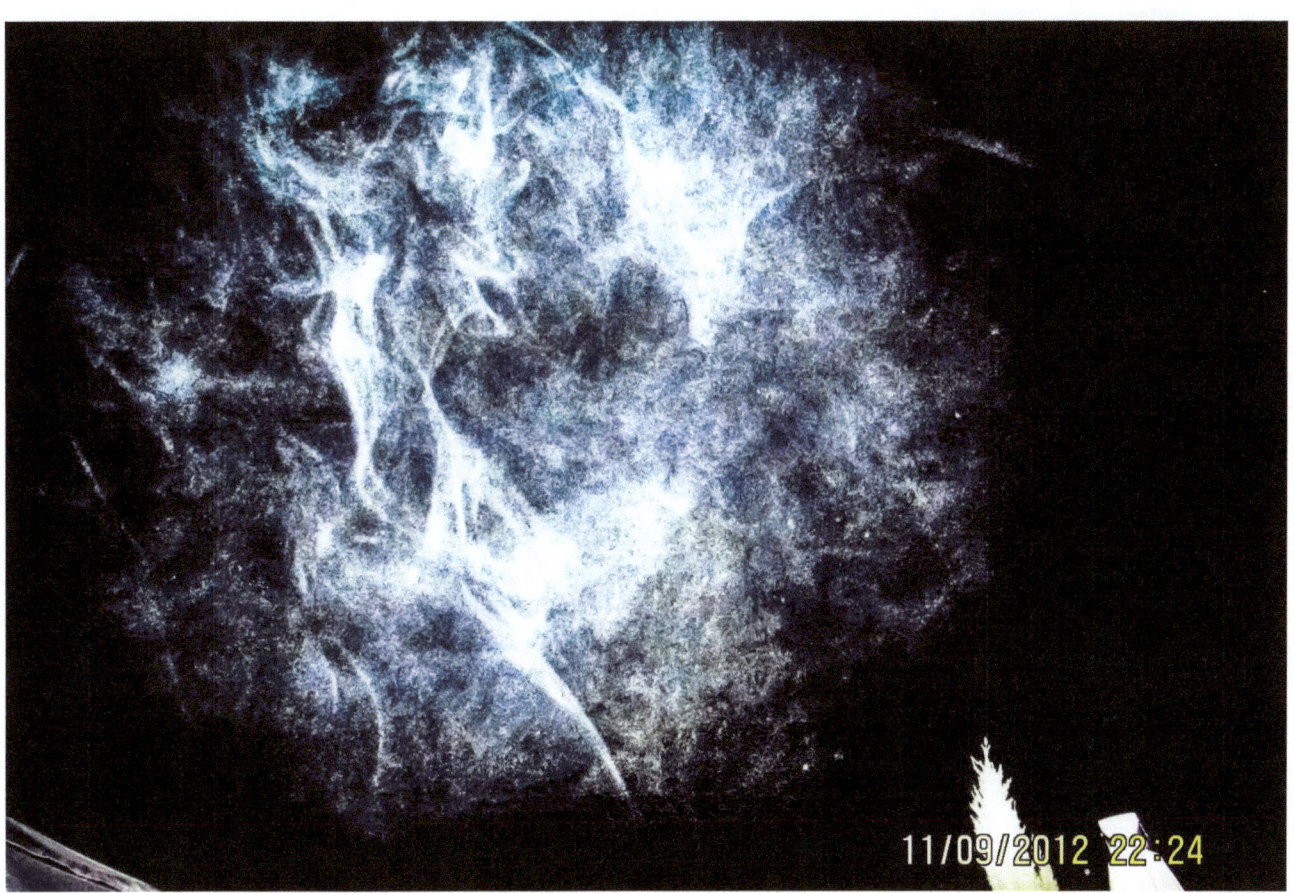

A Sonoma County Phenomenon

Photo 45 (top) and Photo 46 (bottom)

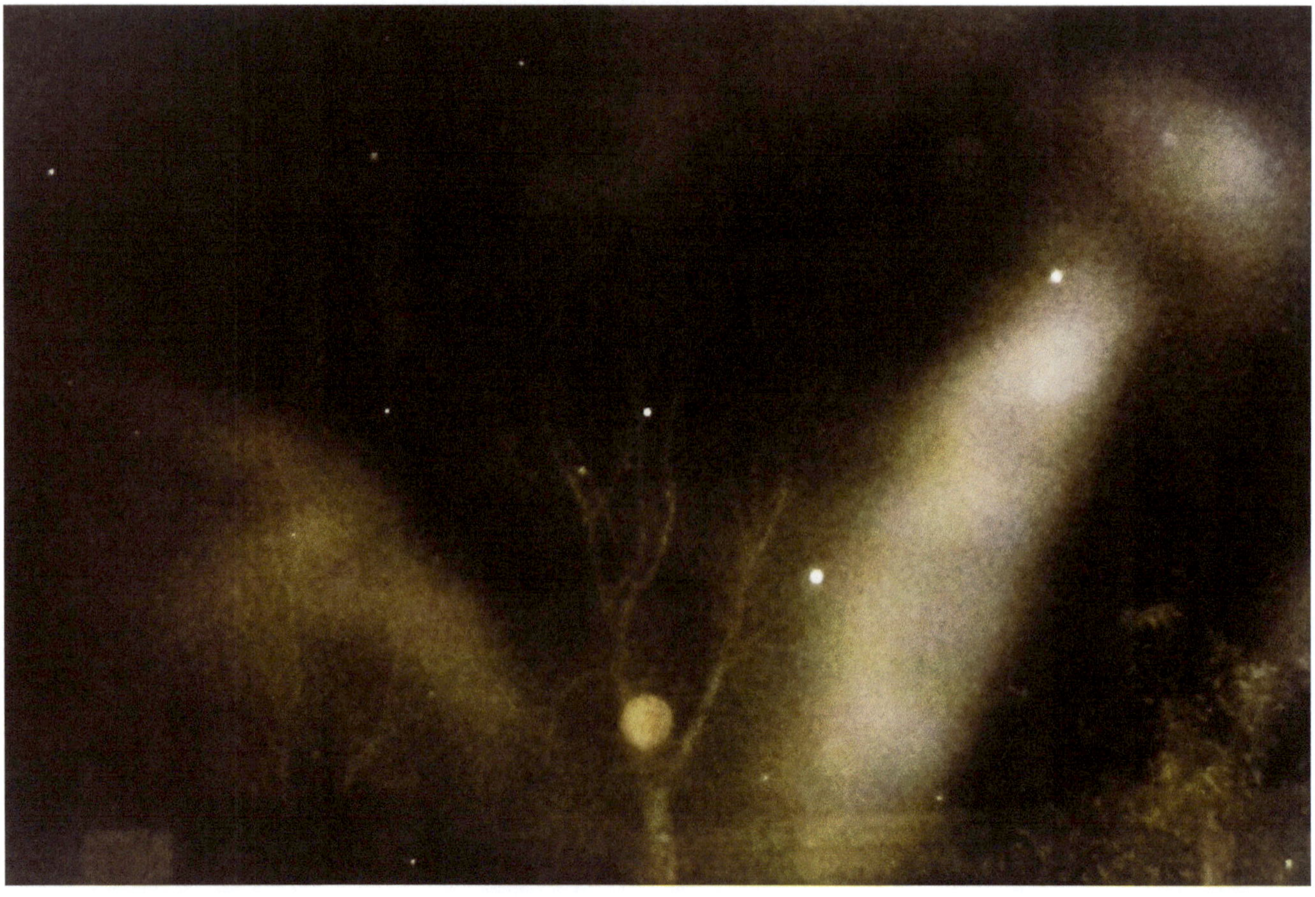

Alien Craft

The photo following is one of the most amazing pictures that Natalie has taken as it appears to show multiple interdimensional craft of various shapes and sizes, each with lights on them. I enlarged five different objects in the photo and each is unique and perplexing. Are these craft moving through the possible worm hole located above Natalie's home?

Photo #47

A Sonoma County Phenomenon

Photo #47

Above: Enlargement of object at center right left of photo.

Right: Enlargement of small object at the left of the photo. It appears to have an energy field around it.

Photo #47 additional enlargements showing detail of four more objects which look like vehicular structures with lights:

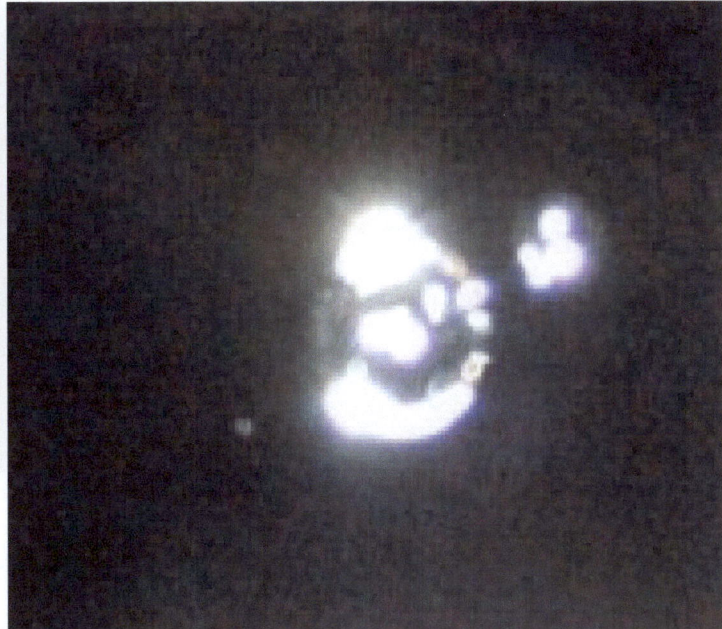
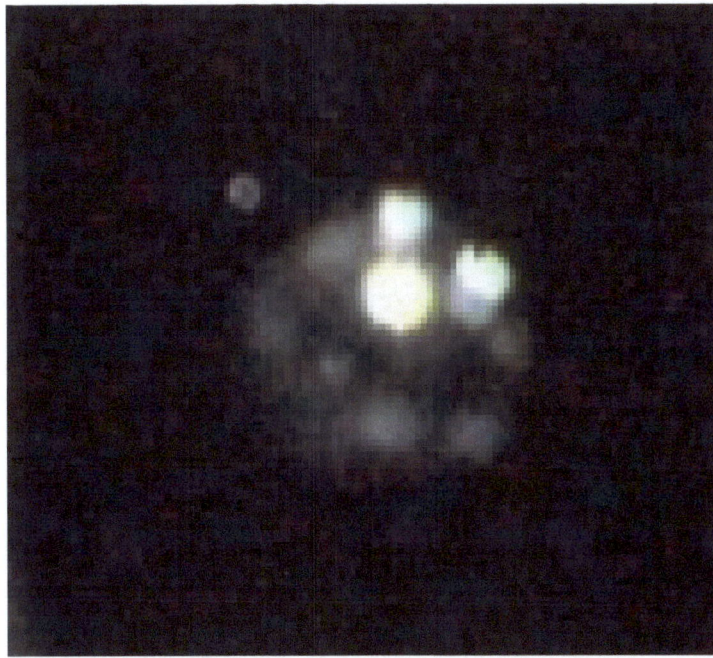
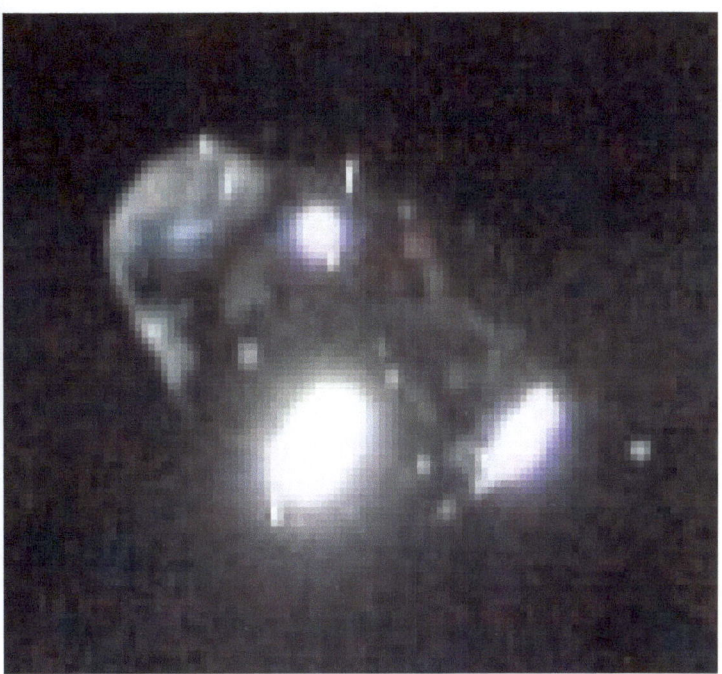

Photo# 48

Taken in the back yard at night. These objects were not visible to the naked eye, but Natalie sensed their presence before taking pictures.

(right) Enlargement of upper section of photos

The Planets and Suns

I've named this section The Planets and Suns—simply because that is what these object look like. Are they really miniature planets passing through a worm hole? What is striking is that many of the objects are well-defined.

Photo 50 (enlargements below)

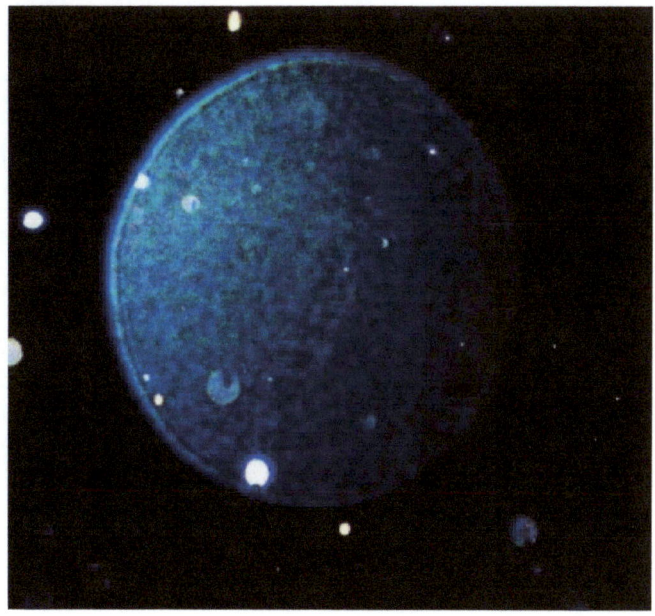

Photo 51:

This picture seems to have it all—planets, craft, orbs, and apparitions.

Enlargement (right)

Photo 52: Enlargement (right)

A Sonoma County Phenomenon

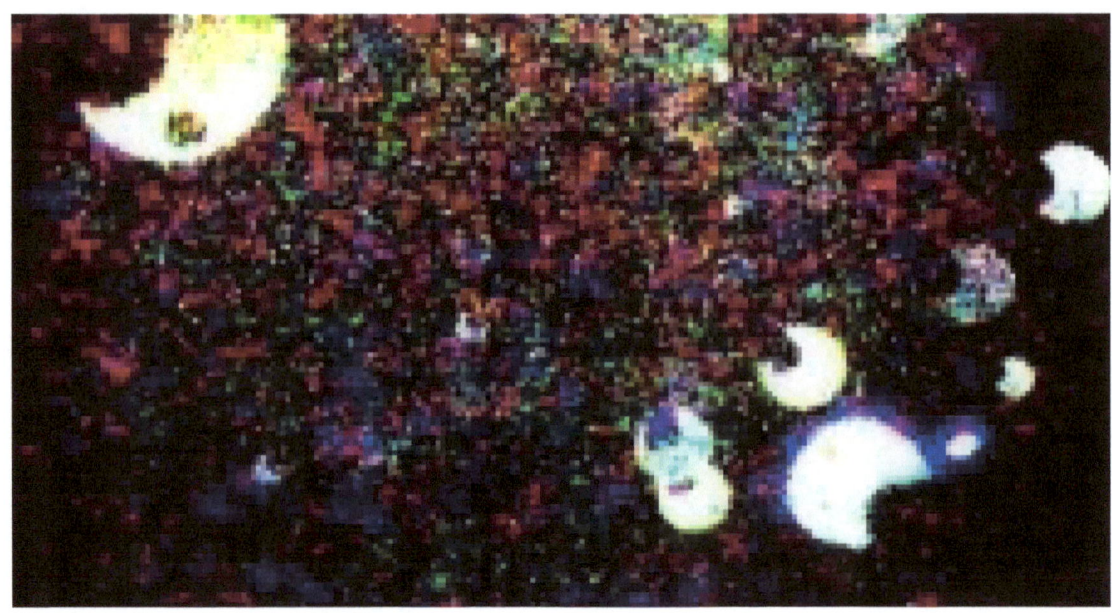

Photo 53: Enlargement of upper center (right)

Margie Kay

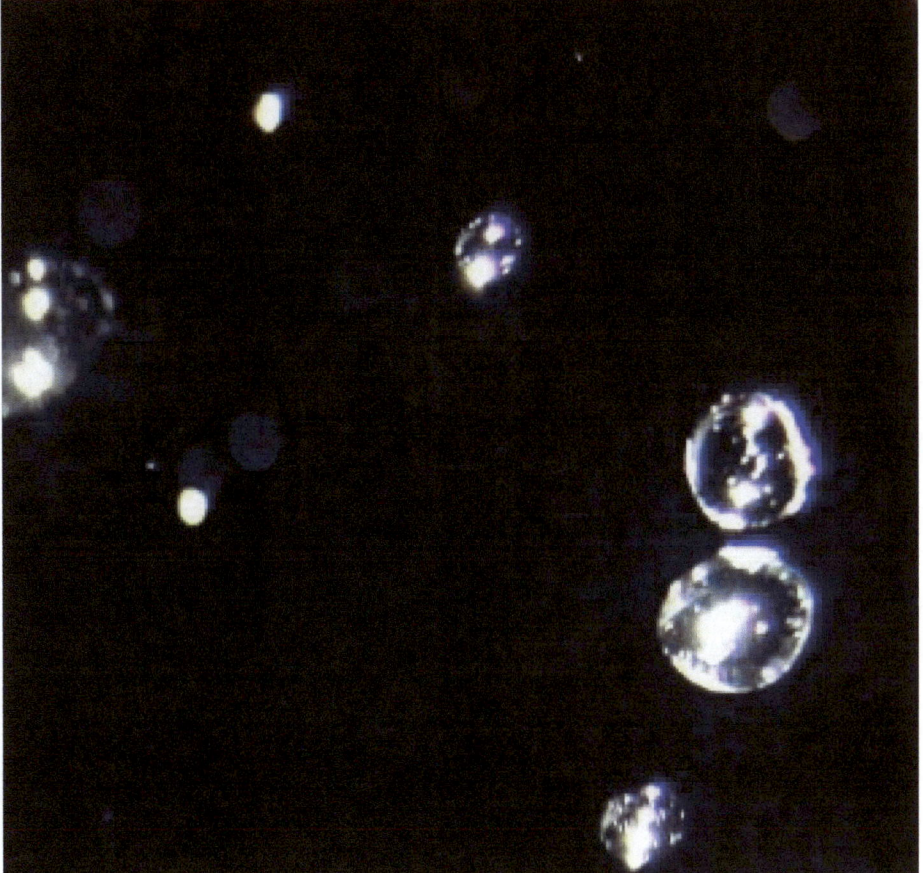

Photo 54:
Enlargement of upper right
(right)

73

A Sonoma County Phenomenon

Photo 55:

Some shapes keep appearing often, including the one at the right that looks similar to a keyhole. Are these rods?

Photo 56:

Brightened to show more detail

Photo 57:

Sometimes the objects look like bright suns, others look line microscopic organisms.

Photo 58:(above)   Enlargement of upper left corner below. Could these be space jellyfish?

Photo 59: Orbs Inside the house. These were seen with the naked eye, then Natalie snapped a picture and captured these objects. She and her family have seen orbs inside the house on a number of occasions. If this area is a portal It would make sense that the objects can pass through solid objects such as doors and windows.

Natalie's Favorites

The following photographs are some of Natale's favorites. She feels that she had a strong communication with whatever intelligence is behind the objects.

Photo #60: Taken at dusk under clear skies. The orbs were visible to the naked eye and captured on camera.

Photo #61: Taken at dusk under clear skies. The orbs were visible to the naked eye and captured on camera.

Photo #62:

Natalie said that this photo is one of Jacques Vallee's favorites since she saw this phenomenon with the naked eye prior to taking the photographs. At first she thought it was a Redwood tree. She took 20 photos of this object, which moved around slightly. She calls this "the gateway."

Photo #63: Natalie has named the above "Blue Monster"   Photo #64 below: Lady with a Hat

A Sonoma County Phenomenon

Photo #65
Natalie named this "Lion Ghost"

Photo #66:
(Below)
Witch Flying By

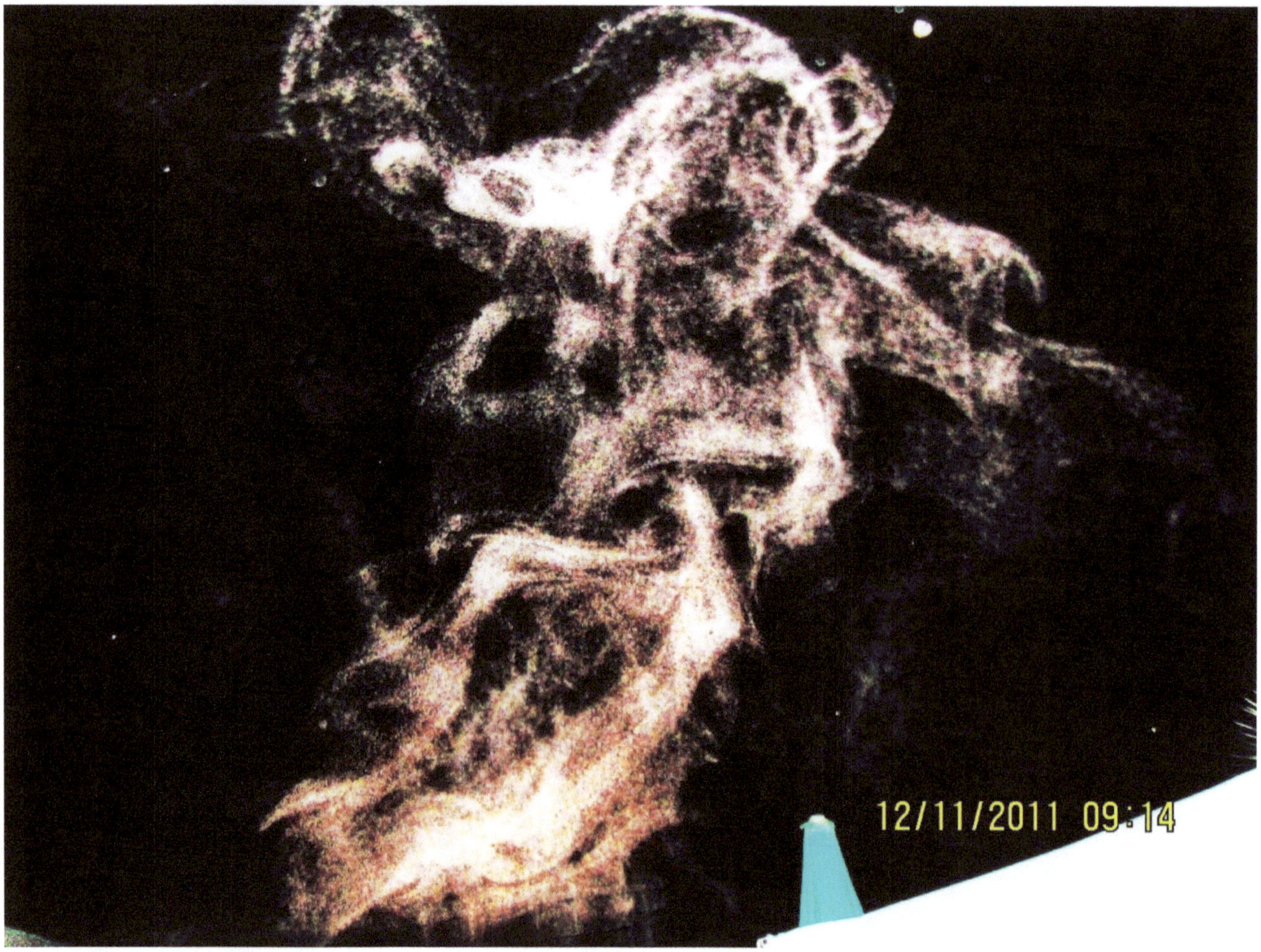

Photo #67

Natalie named this "White Image"

Photo #68: (Below) "Big Mouth Ghost"

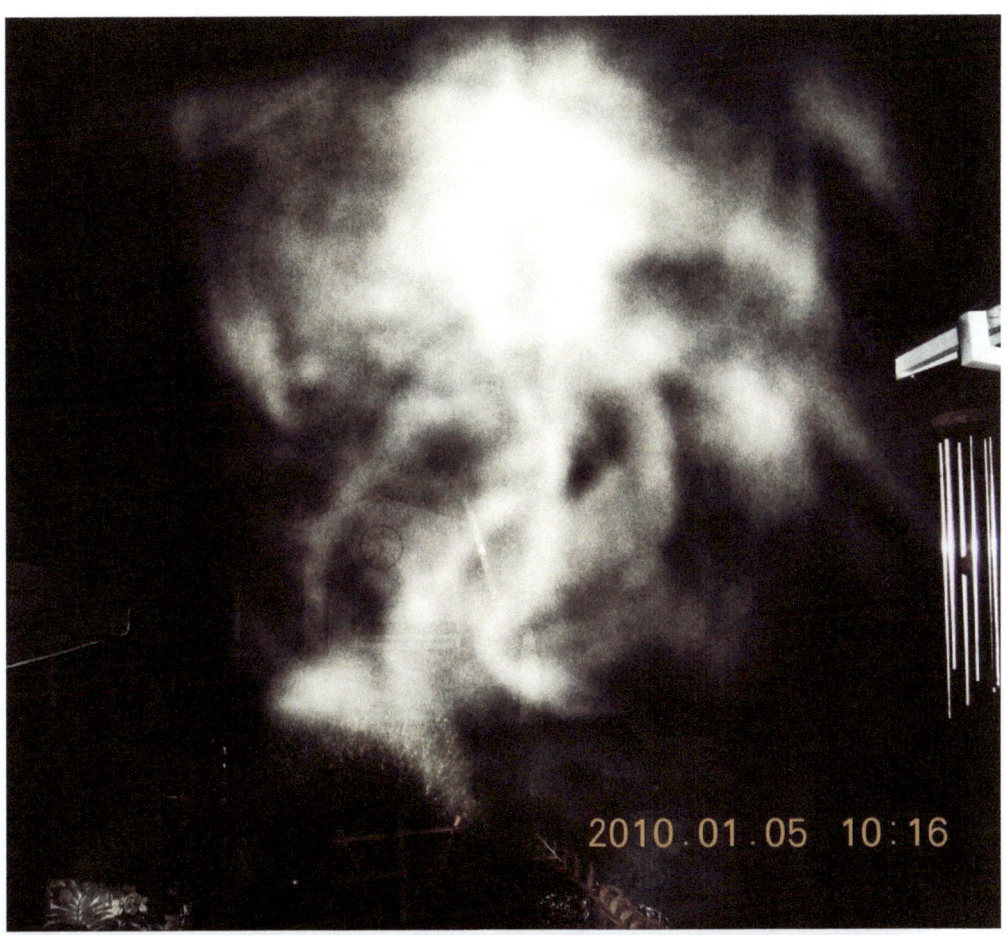

A Sonoma County Phenomenon

Photo #69: Possible craft

Photo #70 and #71 above:
Natalie feels that these are some type of vessels.

Photo #72: (Right)
Natalie calls this "Space Dog."

84

Photo #73:

These objects look like Nebula or craft.

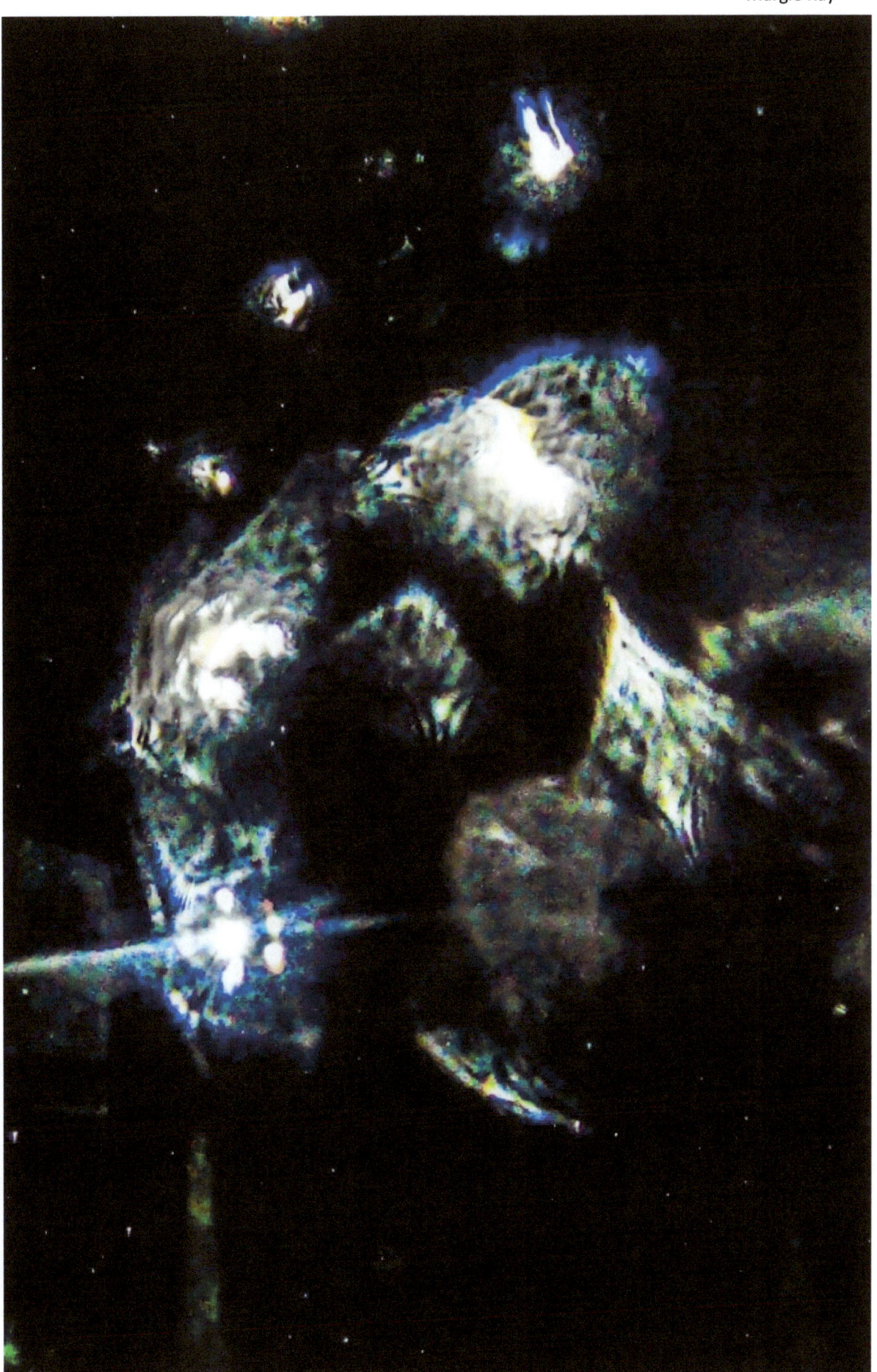

From Natalie: "Here is a special set of images I got on May the 4th, 2015 in a freak rain downpour. The rains had stopped earlier, but returned on this night. I had just started taking photos during the rains thinking that nothing would show up because I thought that nothing could show up with it raining. How wrong I was. his was the beginning of me taking photos during the winter months. It was 4 years into this when I finally realized how wrong I was and how completely the opposite it was. These images came together shot by shot and look at how the image that looks to me as a ship or vessel of some sort changes in every shot.   It's up in the upper left corner and the bubble in one shot looks to me like an alien form."

Photo #74

Photo #75

Photo #76

Photo #77

Photo #78

Photo #79

Photo #80: Continued from previous page.
Right: Enlargement of object in lower left corner of photo.

From Natalie: "This is the ghost tree, one of my favorites, there and then gone. This is a very beautiful tree with perhaps having all the knowledge of the universe. When I first got this shot I had to think for a second about if there was a tree at this location in my yard or not and, of course, there was not. I would catch a glimpse of a large shadow and hear a screeching sound, when finally one day I captured this photo of the object."

Photo #81: Ghostly Tree

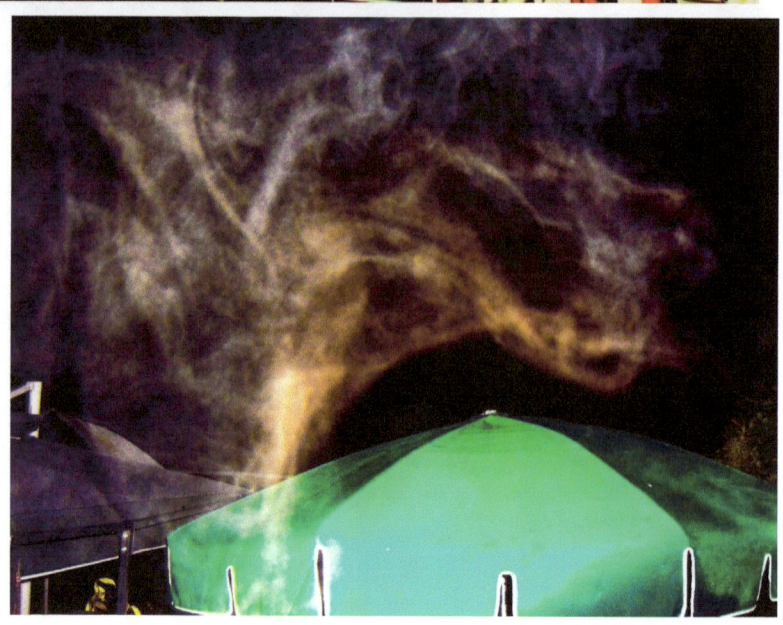

Photo #82:

This is what Natalie calls The snake image. She believes that a serpent or snake occupies her back yard area. This type of object appears in many of her photographs.

Photo #83: The Blue Boy Ghost.

From Natalie: "This little boy is precious and looks kind of sad maybe. This image was used by Grub Street at Halloween for contestants to write a short story about it. It made me proud."

Photo #84:

The Dog Face Ghost. Natalie did not see anything with the naked eye at the time this photo was taken, but she sensed something was there so she snapped this picture.

Photo #85:

The Bat Wings. Natalie heard a screeching sound then saw a huge black shadow and snapped this picture.

Alien Hands?

Photo #86:
Natalie believes that the photo at the right may be a pair of alien hands She enlarged a portion of a larger photo. She says "How wonderful this photo is, and who knows for sure what it is exactly.:

Photo #87: (below)

Alien 2

A Sonoma County Phenomenon

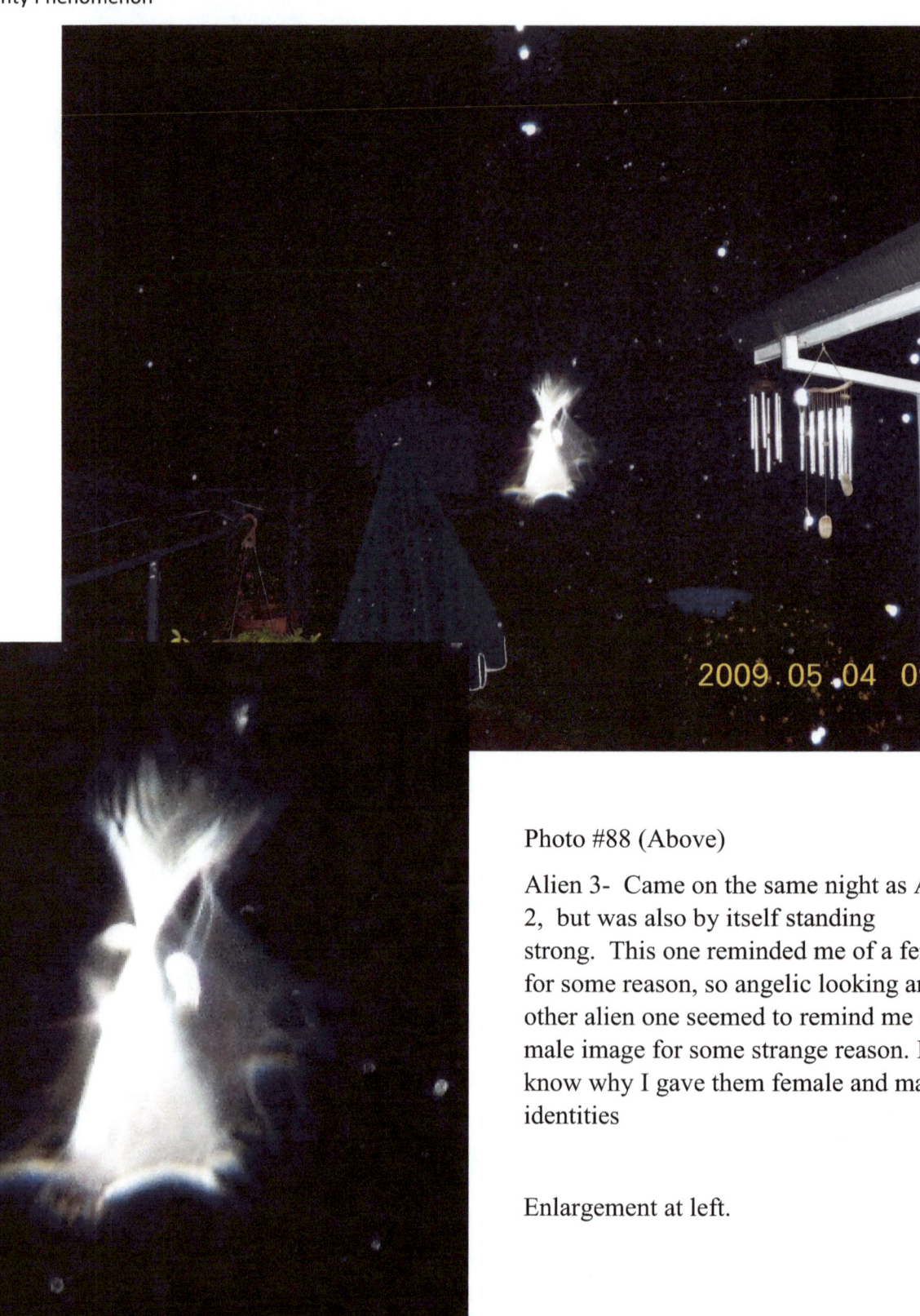

Photo #88 (Above)

Alien 3- Came on the same night as Alien 2, but was also by itself standing strong. This one reminded me of a female for some reason, so angelic looking and the other alien one seemed to remind me of a male image for some strange reason. I don't know why I gave them female and male identities

Enlargement at left.

Following are lower resolution photos taken by Natalie during her initial discovery of the phenomena:

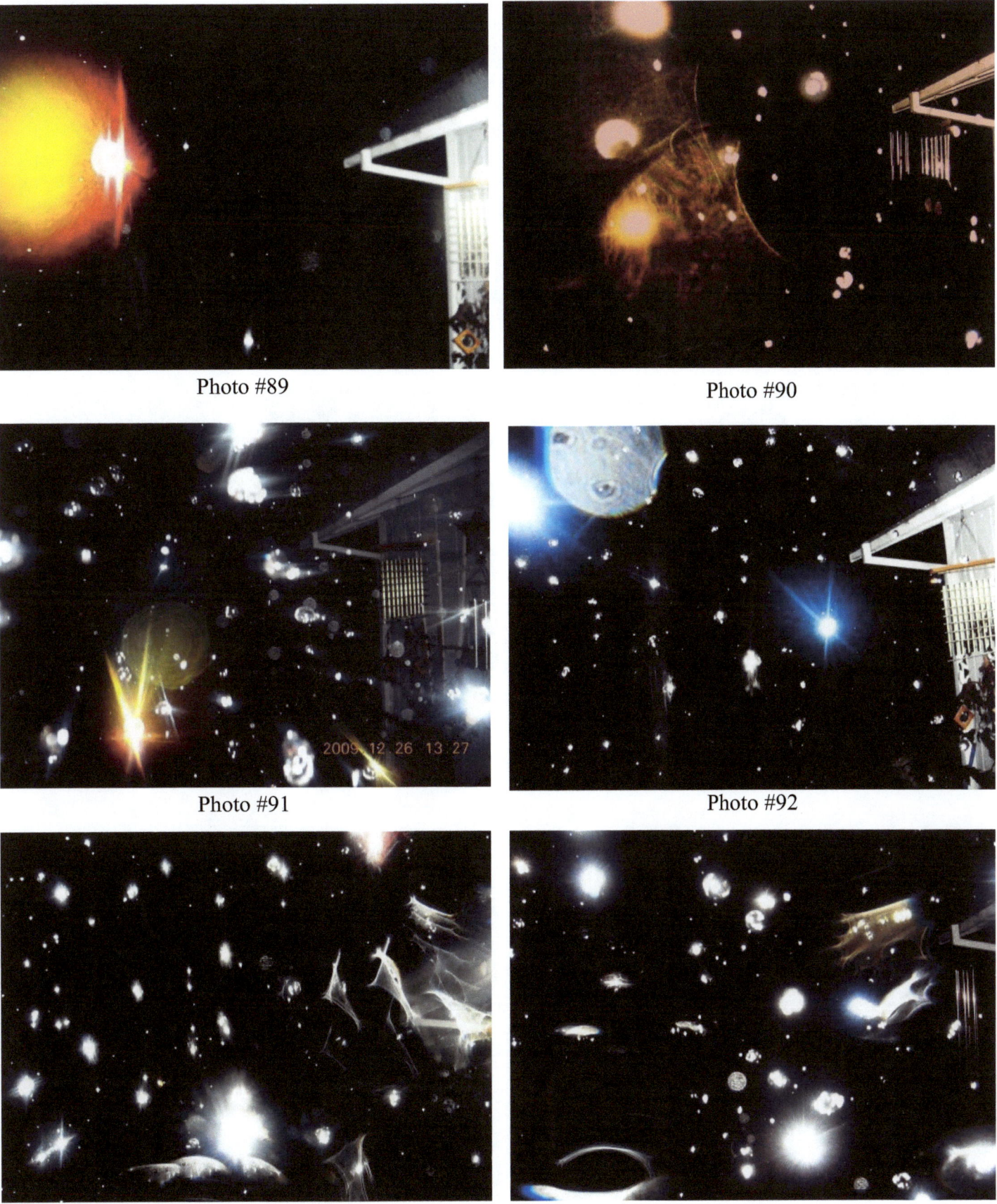

Photo #89

Photo #90

Photo #91

Photo #92

Photo #93

Photo #94

Photo #95

Photo #96

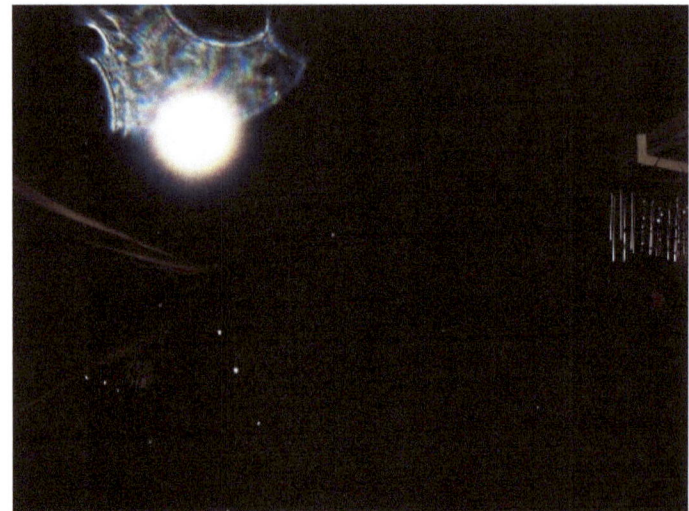
Photo #97

Photo #98

Photo #93

Photo #100

Photo 100 enlarged (above) and close up of upper right corner (right)

A Sonoma County Phenomenon

Detail of Photo 100

Close up of upper right section, slightly above the orange object

Below: Close up of upper center portion of photo

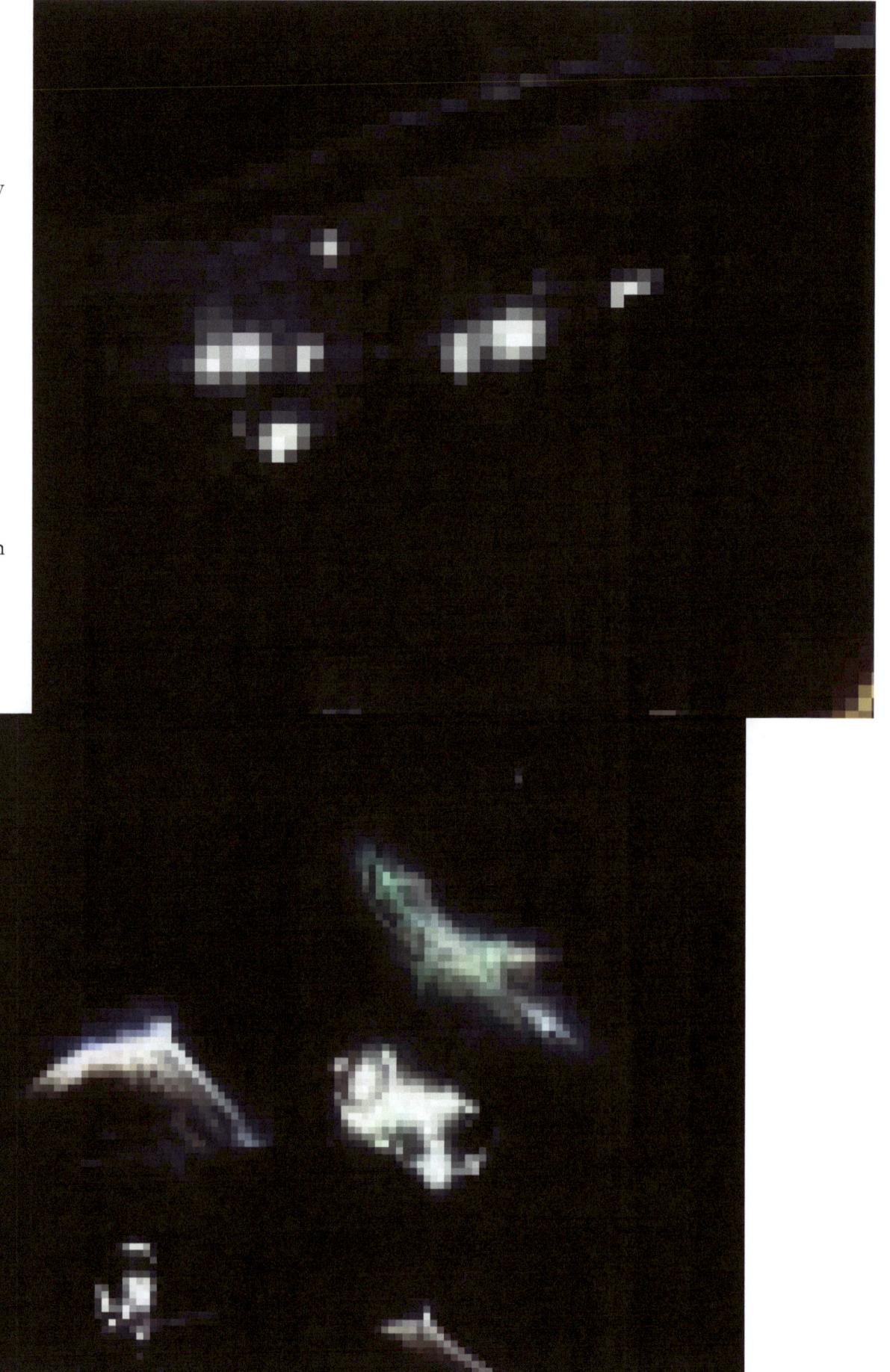

Margie Kay

Our Multi-Dimensional Universe

Wormholes are theoretical shortcuts between distant regions of space and time that are
predicted by the theory of general relativity, first proposed by Albert Einstein and Nathan Rosen in 1935.

A **wormhole** (or **Einstein–Rosen bridge** or **Einstein–Rosen wormhole**) is a speculative structure linking points in spacetime. A wormhole can be visualized as a tunnel with two ends at separate points in spacetime (i.e., different locations, or different points in time, or both.)

Wormholes are consistent with the general theory of relativity, but whether wormholes actually exist remains to be seen. Many scientists postulate wormholes are merely a projection of the 4th dimension, analogous to how a 2D being could experience only part of a 3D object.

In my observations using remote viewing, what I see at this particular location in Sonoma County is something that is similarly shaped to the typical wormhole drawing with two large ends far out into space which narrow significantly in the center, and this central tube goes right through Natalie's yard into the Earth and out the other side of the planet.

Theoretical Physicist Dr. Michio Kaku states in his YouTube video The Universe Has 11

Albert Einstein during a lecture in Vienna in 1921

Dimensions—Think Big: "We believe, though we cannot yet prove, that our multiverse of universes is 11-dimensional. So think of this 11-dimensional arena and in this arena there are bubbles,
bubbles that float and the skin of the bubble represents an entire universe, so we're like flies trapped on fly paper.

We're on the skin of a bubble. It's a three dimensional bubble. The three dimensional bubble is expanding and that is called the Big Bang theory and sometimes these bub-

Big Bang. So we even have a theory of the Big Bang itself."

There are inter-dimensional aspects to this site as well, as I've observed beings or craft coming in and out of the wormhole while remote viewing. Think of these objects traveling through a highway of sorts, then something catches their attention (in this case, Natalie) then they move off the highway and into the yard, then back to the highway. Hopefully, this will be scientifically proven someday soon.

bles can bump into each other, sometimes they can split apart and that we think is the

Conclusion

By now the reader is likely as perplexed as the author. Its difficult to draw a definitive conclusion about the extremely unusual events that have occurred at this location, or decipher the photos, so one can only speculate. In order to draw a reasonable conclusion I had to consider the facts:

1. Natalie seems to have the ability to call these objects, beings, or craft telepathically and ask them to appear. If this is true, then the objects either have a consciousness themselves or are craft operated by a sentient being.

2. Many of the objects are visible to the naked eye and have been seen by others, so this eliminates complications from camera equipment.

3. Natalie has taken over 250,000 photographs of the phenomena. For someone to manipulate that many photos would in my mind be nearly an impossible feat, so a hoax is not likely. Examination of some of the photos by an expert showed no manipulation.

4. Physical evidence such as water turning on of its own accord, unexplained burn marks on the ground, high EMF readings and temperature differences at the site cannot be disputed or explained.

5. Scientists and investigators have been keenly interested in this location, indicating that there must be good reason for their interest.

6. Other people have been able to capture photographic images of anomalous objects as well, which indicates that the phenomena exists with or without Natalie's presence, although the objects seem to have become aware of her and react.

Natalie feels that whatever is at her site is something that has been there forever and will likely always exist. She believes that the objects or beings have a consciousness since they seem to react to her request to appear.

It is my opinion that there are a combination of factors involved which contribute to the phenomena including the location of the site, the ley lines and intersecting coordinates which are likely creating a portal or worm hole, and that the number of artifacts captured in Natalie's photos are directly related to her telepathic connection to the consciousness involved. Natalie may also be attracting more of these objects to her than would normally exist.

Whether the worm hole or portal opens and closes at particular times remains to be seen, however, Natalie says that the area is most active in the winter and around the time that storms occur. I find this interesting, since I've found a connection to UAP (Unidentified Ariel Phenomena) and the increase of number of these objects reported right after a major storm. I've seen anomalous objects myself after storms occur in different locations around Kansas City. Perhaps negative IONs in the air create the framework for better visibility of these objects.

After knowing Natalie for a number of years and carefully reviewing her photographs, I believe she is telling the truth about everything she has experienced and everything she has seen both with and without the use of a camera. Natalie just wants answers, and so do I.

This location may eventually help to prove the existence of worm holes and an inter-dimensional gateway as scientists and investigators continue to investigate this location and others like it around the world.

Author's Note: Natalie will not allow anyone other than the investigators currently working on the case to visit her site. She is already overwhelmed, so please do not attempt to contact her. All inquiries should go to the author.

Sonoma County map from Google Earth

Photo: Sonoma County California by Adobestock.com

References and Suggested Reading

Natalie Roberts' website: www.asonmacountyphenomenon.com

Ley Lines: www.wikipedia.com

SpaceX Launch: https://www.popsci.com/spacex-launch-look-strange-alien

Ray Parkes blog post "California Ley Lines" https://rayparkes.com/california-ley-lines

Paranormal Activity on Hiking Trails: https://backpackerverse.com/10-sonoma-county-hiking-trails-with-insane-paranormal-activity/5_West_County_Regional_TrailSebastopol.

Gravity Hill: https://www.onlyinyourstate.com/northern-california/strange-phenomenon-norcal/

Anti-Gravity and the World Grid by David Hatcher Childress, (Adventures Unlimited Press) Paperback, June 1, 1987

MUFON (Mutual UFO Network): www.mufon.com

The Kansas City UFO Flaps by Margie Kay www.margiekay.com

Missouri MUFON: www.missourimufon.org

Sonoma County, CA Wikipedia: https://en.wikipedia.org/wiki/Sonoma_County,_California

The Universe Has 11 Dimensions—Think Big: by Michio Kaku https://youtu.be/jI50HN0Kshg

About the Author

Margie Kay is a veteran ufologist and professional paranormal researcher, having completed over 1,200 investigations of UFO sightings, close encounters, and paranormal events. Kay is the editor and owner of Un-X Media Publishing and author of multiple books and articles.

She was the publisher and editor of *Un-X News Magazine* from 2001-2015 and host of *Un-X News Radio show on the KGRA Network* and *Quest Radio Show* on 1140AM Kansas City. Margie has been the director of the Mysteries of the Universe Conference since 2007. She is the Director of Quest Paranormal Investigations based in Kansas City, Missouri.

Margie owns a forensic investigation company and is an ex-private investigator. She is an accomplished musician, remote-viewer and psychic medium. She has helped to solve over 60 missing person, homicide, and theft cases for law enforcement and private investigators. Kay uses her RV skills on paranormal investigation cases.

Kay is a national speaker and instructor. She teaches remote viewing and psychic skills techniques for investigators, and speaks at conventions and meetings nationwide about her own encounters and the investigations she has worked on over the years.

Margie lives in Independence, Missouri with her husband, Geno, and their cat, Patches. Margie's daughter is Maria Christine, who is an author of paranormal romance books, and works as Margie's finance manager.

Contact:

Email: margiekay06@yahoo.com
editor@unxmedia.com
816-833-1602

Websites:

www.margiekay.com

www.unxmedia.com

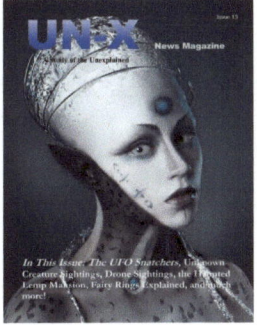

Publications by Un-X Media

Family Secrets 2017 by Jean Walker

Haunted Independence 2013 by Margie Kay

Gateway to the Dead: A Ghost Hunter's Field Guide 2013 by Margie Kay

The Kansas City UFO Flaps 2017 by Margie Kay

Unexplained Missouri 2020 by Margie Kay

The Remote Viewing Workbook 2019 by Margie Kay

A Sonoma County Phenomenon 2020 by Margie Kay

Un-X News Magazine 2001-2015

The Color Therapy Wall Chart 1999

Rules for Goddesses by Margie Kay 2003

The Fast Movers by Margie Kay, Bill Spicer, and Wayne Lawrence 2020

Doorway to Spirit by Devin Listrom 2020

More books coming soon!

www.unxmedia.com

All books available at Amazon.com and BarnesandNoble.com

UNXMEDIA

PUBLISHING

www.ingramcontent.com/pod-product-compliance
Lightning Source LLC
Chambersburg PA
CBHW042000150426
43194CB00002B/72